Progress in Mathematics
Vol. 19

Edited by
J. Coates and
S. Helgason

Birkhäuser
Boston · Basel · Stuttgart

Takayuki Oda

Periods of Hilbert Modular Surfaces

1982

Birkhäuser
Boston • Basel • Stuttgart

Author:

Takayuki Oda
Department of Mathematics
Hokkaido University
Sapporo, 060
Japan

Library of Congress Cataloging in Publication Data

Oda, Takayuki, 1950-
Periods of Hilbert modular surfaces.
(Progress in mathematics ; v. 19)
Bibliography: p.
1. Forms, Modular. 2. Hilbert modular surfaces.
I. Title. II. Series: Progress in mathematics
(Cambridge, Mass.) ; 19.
QA243.03 512'.7 82-4315
ISBN 3-7643-3084-8 AACR2

CIP-Kurztitelaufnahme der Deutschen Bibliothek

Oda, Takayuki:
Periods of Hilbert modular surfaces / Takayuki Oda.
Boston ; Basel ; Stuttgart : Birkhäuser, 1982.
(Progress in mathematics ; Vol. 19)
ISBN 3-7643-3084-8

NE: GT

ISBN 3-7643-3084-8
Printed in USA

Foreword and acknowledgments.

The arithmetic theory of periods of modular forms is revealing its nature as Diophantine index theorem. This paper is an attempt to amplify this universal principle by discussing a special case: Hodge structures of Hilbert modular surfaces.

The first draft of this paper was written during my stay at Harvard University from October 1980 to September 1981, supported by the Japan-U.S. Exchange Fund, Harvard University.

I express my thanks to Professor Barry Mazur, who shared the belief of the main theorems of this paper, when they were yet conjectures, and also to Professor David Kazhdan for his pertinent comments on this paper and other related works. I owe Professor Michio Kuga for some remarks on abelian varieties attached to K3 surfaces. Professors Tetsuji Shioda and Yasuo Morita have been encouraging me constantly.

This paper is traced back to a short note written several years ago when I was a graduate student, inspired by the preprints of Manin and Deligne shown by Professor Yasutaka Ihara. I thank him cordially for his encouragement and patience.

It is a consolation for me to find that the idea of the late Professor Takuro Shintani plays an important role in this paper, and to recollect his individualities.

Sapporo, 29 January 1982. Takayuki Oda

Table of contents

Chapter 0. Historical background, motivations and outline of the contents.

As the readers might be unfamiliar with the rather special topics dealt in this book, the author would like to explain the background and the motivations of our subject before summarizing the main results.

i) Tate conjecture. The Tate conjecture on algebraic cycles is one of the important problems in Diophantine Geometry, which predicts the following (cf. Tate [61]).

Let X be a smooth proper geometrically connected scheme over an algebraic number field k, and let $H^i(X\times\bar{k},\mathbb{Q}_\ell)$ be the i-th ℓ-adic étale cohomology group of X, where $\bar{k}$ is the separable closure of k. Let $A^d(X)$ be the free group generated by the irreducible subvarieties Y of X of codimension d defined over $\bar{k}$, and let

$$c_\ell^d : A^d(X) \longrightarrow H^{2d}(X\times\bar{k},\mathbb{Q}_\ell)(d)$$

be the ℓ-adic cycle mapping, where $H^{2d}(X\times\bar{k},\mathbb{Q}_\ell)(d)$ is the d-ple Tate twist of $H^{2d}(X\times\bar{k},\mathbb{Q}_\ell)$.

X has good reductions at almost all prime ideal p of k. For each good prime p, we can define the characteristic polynomials $P_p^i(t)$ of the geometric Frobenius $F_p^i : H^i(X\times\bar{k},\mathbb{Q}_\ell) \longrightarrow H^i(X\times\bar{k},\mathbb{Q}_\ell)$ induced from the Frobenius mapping $F_p : X(p) \longrightarrow X(p)$ by the specialization theorem of Grothendieck: $H^i(X\times\bar{k},\mathbb{Q}_\ell) \xrightarrow{\sim} H^i(X(p)\times\overline{k(p)},\mathbb{Q}_\ell)$. Here $X(p)$ is the reduction mod p of X at p and $k(p)$ is the residue field at prime p. The i-th Hasse-Weil L-function $L(s,H^i(X))$ of X is defined as the Euler product

$$L(s,H^i(X)) = \prod_{p;\text{good primes}} P_p^i(Np^{-s})$$

for each i, where Np is the norm of the prime ideal p.

Assume that the image $A_\ell^d(X)$ of the cycle mapping c_ℓ^d has a basis consisting of the images $c_\ell^d(Y)$ of irreducible subvarieties Y defined over k. Then the conjecture claims that the L-function $L(s,H^{2d}(X))$ (which is expected to be continued holomorphically to the whole complex s-plane) has a pole of order $\rho_d(X)=\mathrm{rank}_{\mathbb{Q}_\ell} A_\ell^d(X)$ at s=d+1.

Up to now we scarcely have any general results except some chosen examples. I refer to two of them here. In both cases X is a surface and d=1. In the first case when X is an elliptic modular surface, Shioda [60] has determined the Picard number $\rho_1(X)$ of X, and the result of Jacquet-Shalika [57] on the non-vanishing of the special values of L-functions implies that the order of the poles of $L(s,H^2(X))$ at s=2 is as expected for any abelain extension k of $\mathbb{Q}$. For another case when X is a product of two elliptic curves defined over $\mathbb{Q}$ which are factors of the jacobian varieties of modular curves, this conjecture is checked by Ogg [34] for k=$\mathbb{Q}$.

For the time being we may have few examples of algebraic varieties whose Hasse-Weil L-functions would be under our control, except modular varieties. Therefore, it seems an interesting problem to check the Tate conjecture for Hilbert modular surfaces.

ii) Periods of modular forms. In a series of papers (cf. [16], [17], [19], [56]), Hirzebruch and Zagier have constructed certain algebraic cycles on Hilbert modular surfaces by using the modular embeddings of modular curves, and investigated them in detail.

There are some reasons to believe (and in fact we can show) that they have constructed sufficiently many algebraic cycles which generate $A_\ell^1(X)$ for some X. To show this fact (i.e. to determine the Picard number $\rho_1(X)$ of X), we have to find some sorites to exculde the possibility of algebraic cycles on X more than construceted by Hirzebruch-Zagier.

The Lefschetz criterion of algebraic cycles on surfaces tells that for any proper smooth algebraic surface X over $\mathbb{C}$, a 2-cycle γ of $H^2(X(\mathbb{C})^{an},\mathbb{Z})$ is an algebraic cycle if and only if the period integrals

$$\int_\gamma \omega = 0$$

for any holomorphic 2-form ω on X.
Thus we are naturally led to consider the periods of Hilbert modular cusp forms of weight 2, and to represent these periods as special values of L-functions attached to modular forms to show non-vanishing of them.

From this point of view, one of our subjects is an attempt of generalization of the theory of modular symbols, developped by Shimura [41], Birch [6], Manin [22], Mazur [23], and Shimura [47], [48].

The theory of periods of elliptic modular forms was started by Shimura [41]. In relation to the conjectures of Birch & Swinnerton-Dyer [5] and Weil uniformization, Manin [22] and Mazur [23] systematically took up this theory anew.

In Deligne [12], we find general conjectures on the relations of Hodge structures and special values of L-functions, which contain the conjecture of Birch & Swinnerton-Dyer as a special case. We investigate this type of realtions for Hilbert modular surfaces. However, in order to push forward our problem to determine the Picard numbers of Hilbert modular surfaces, we need another viewpoint.

iii) Weil uniformization. The Taniyama-Weil conjecture or Langlands philosophy says the following.

For any elliptic curve E defined over an algebraic number field k, there exists an automorphic form f_E over $GL_2(A_k)$ such that

$$L(s,f_E)=L(s,H^1(E)) \quad \text{modulo finite number of Euler factors.}$$

Here A_k is the adelization of k, and $L(s,f_E)$ the L-function attached to f_E.

As is well-known, this conjecture has the following stronger geometric version for $k=\mathbb{Q}$ (cf. Weil [54], Mazur & Swinnerton-Dyer [23]).

Let C_N be the smooth projective model over $\mathbb{Q}$ of the modular function field $\mathbb{Q}(j(\tau), j(N\tau))$, where $j(\tau)$ is given by

$$j(\tau)=1728\,\frac{g_2^3(\tau)}{g_2^3(\tau)-27g_3^2(\tau)},$$

where τ is a point of the complex upper half plane H, and $g_2(\tau)$ and $g_3(\tau)$ are Weierstrass's functions.

Conjecture of Weil uniformization. For any elliptic curve E over $\mathbb{Q}$, there are a modular curve C_N and a surjective morphism $\varphi_E : C_N \longrightarrow E$ defined over $\mathbb{Q}$.

Let us consider how to generalize this geometric version for totally real number fields F. For simplicity, we assume that the class number of F is 1, and moreover that $[E_F:E_F^+]=2^{g-1}$, where E_F and E_F^+ are the group of units and the group of the totally positive units of F, respectively, and $g=[F;\mathbb{Q}]$.

The group $SL_2(F)$ acts on the product H^g of g copies of the complex upper half plane H in the usual manner. For any ideal $\mathfrak{n}$ of the integer ring O_F of F, we denote by $V_{\mathfrak{n}}$ the quotient modular variety $\Gamma_0(\mathfrak{n})\backslash H^g$, where $\Gamma_0(\mathfrak{n})$ is a subgroup of $SL_2(F)$ given by

$$\Gamma_0(n)=\{ \begin{pmatrix}\alpha & \beta \\ \gamma & \delta\end{pmatrix} \in SL_2(O_F) \mid \gamma \equiv 0 \bmod n \}.$$

Let $H^g(V_n,\mathbb{Q})$ be the g-th rational cohomology group of V_n which has a mixed Hodge structure by Deilgne [11]. Let $p_\infty(F)$ be the set of all embeddings of F into $\mathbb{C}$. Then the cardinality of $p_\infty(F)$ is g.

As a generalization of Weil uniformization, we propose the following Conjecture (Hodge realization). For any elliptic curve E over F, there exist a modular variety V_n for some ideal n of O_F, and a monomorphism of Hodge structures

$$\varphi_E^*: \bigotimes_{\tau \in p_\infty(F)} H^1(E^\tau(C),\mathbb{Q}) \hookrightarrow H^g(V_n,\mathbb{Q}),$$

where E^τ is the extension of scalars of E with respect to the embedding $\tau: F \longrightarrow \mathbb{C}$, and $E^\tau(\mathbb{C})$ is the analytic variety associated to its $\mathbb{C}$-valued points.

iv) Main Conjecture A^{split}. The previous conjecture leads us to the following Main Conjecture A^{split}.

Let $\{W.\}$ be the weight filtration of mixed Hodge structures defined in Deligne [11]. Consider $W_gH^g(V_n,\mathbb{Q})$ which is a homogeneous Hodge structure of weight g. Then, the Hecke operators $T(b)$ act on this Hodge structure naturally. Suppose that f is a cusp form of weight 2 with respect to $\Gamma_0(n)$, which is an eigenform of all Hecke operators and a new form: $T(b)f=a_bf$. Let K_f be the subfield of $\mathbb{C}$ generated by the eigenvalues a_b of f over $\mathbb{Q}$, and let $\phi_f: H \longrightarrow K_f$ be the homomorphism of the Hecke ring H to K_f defined by

$$\phi_f(T(b))=a_b$$

for each $T(b)$.

For each primitive form f of weight 2 with respect to $\Gamma_0(n)$, we put

$$H^g(M_f,\mathbb{Q})=W_gH^g(V_n,\mathbb{Q})\otimes_H K_f.$$

Then $H^g(M_f,\mathbb{Q})$ is a rational homogeneous Hodge structure of weight g with a homomorphism $\theta_f: K_f \hookrightarrow \mathrm{End}(H^g(M_f,\mathbb{Q}))$. Thus $H^g(M_f,\mathbb{Q})$ is a rational Hodge structure with coefficients in K_f (cf. [12]), or a K_f-Hodge structure in our terminology.

Main Conjecture A^{split}. For each primitive form f of weight 2, there exists an abelian variety A_f of dimension $[K_f:\mathbb{Q}]$ defined over F with a homomorphism

$$\Theta_f: K_f \hookrightarrow \mathrm{End}(A_f)\otimes_{\mathbb{Z}}\mathbb{Q}$$

such that there is an isomorphism of K_f-Hodge structures

$$H^g(M_f,\mathbb{Q}) \xrightarrow{\sim} \bigotimes_{\tau\in p_\infty(F)} {}_{K_f} H^1(A_f^\tau(\mathbb{C}),\mathbb{Q}),$$

where A_f^τ is the extension of scalars of A_f with respect to the embedding $\tau: K_f \longrightarrow \mathbb{C}$.

This book is an interim report on our effort to show Main Conjecture A^{split} for real quadratic fields F. We shall discuss the more general cases and other related conjectures in subsequent papers.

In the next section, we explain how this conjecture is related with the determination of the Picard numbers of Hilbert modular surfaces.

Remark. There is a method of attaching abelian varieties to the space of modular forms by means of Weil's intermediate jacobian varieties, developped by Shimura [41], Mountjoy [26] (and especially for Hilbert modular forms by Hida [15]). Our abelian variety A_f is different from these abelian varieties. Our Main Conjecture A^{split} seems to be compatible with the "theory of motives" of Grothendieck-Deligne (cf. [12]). The Hodge structure $H^g(M_f,\mathbb{Q})$ is not of level 1 for $g\geq 2$ (cf.[58]).

v) Outline of the contents. In the first place, we fix a real quadratic field F with discriminant D, which has a unit with negative norm. In Chapter I, similarly as Deligne [12], we develop a formalism which attaches a Hodge structure $H^2(M_f,\mathbb{Q})$ for any Hilbert modular cusp form f of weight 2 with respect to $SL_2(O_F)$, which is a common eigenform of all Hecke operators:In §1, we investigate the homogeneous part $W_2H^2(S,\mathbb{Q})$ of the mixed Hodge structure $H^2(S,\mathbb{Q})$ of the non-compact Hilbert modular surface S. In §2, we decompose $W_2H^2(S,\mathbb{Q})$ into eigenspaces

$$W_2H^2(S,\mathbb{Q})=(\bigoplus_{f\in\Xi} H^2(M_f,\mathbb{Q})) \oplus \mathbb{Q}(-1) \oplus \mathbb{Q}(-1)$$

with respect to the action of the Hecke ring H. Thus we have a polarized Hodge structure $H^2(M_f,\mathbb{Q})$ of weight 2 attached to an eigenform f, which is of rank 4 over the field of eigenvalues K_f of f. In §3, we define certain nonholomorphic involutive automorphisms G_∞ and H_∞ of a Hilbert modular surface S, and decompose the homology group $H_2(M_f,\mathbb{Q})$ into four eigenspaces $H_2(M_f,\mathbb{Q})_{\pm\pm}$ with respect to these involutions. In §4, by using this decomposition, we reformulate the classical Riemann-Hodge period realtion in terms of the Hodge structure $H^2(M_f,\mathbb{Q})$. This relation is used in §6, and in the proof of Main Theorem B (§17).

In Chapter II, starting with the Hodge structure $H^2(M_f,\mathbb{Q})$, we construct two isogeny classes $A_f^1\otimes\mathbb{Q}$ and $A_f^2\otimes\mathbb{Q}$ of abelian varieties A_f^i $(i=1,2)$ with homomorphisms

$$\theta^i : K_f \hookrightarrow \mathrm{End}(A_f^i)\otimes_{\mathbb{Z}}\mathbb{Q},$$

such that there is an isomorphism of K_f-Hodge structures (Main Theorem A, §7)

$$H^2(M_f,\mathbb{Q}) \xrightarrow{\sim} H^1(A_f^1,\mathbb{Q})\otimes_{K_f} H^1(A_f^2,\mathbb{Q}).$$

Let $H^2(M_f,\mathbb{Q})_{alg}$ be the subspace of $H^2(M,\mathbb{Q})$ generated by algebraic cycles. Then, as a corollary of Main Theorem A (cf. §7.6), we have an isomorphism of $\mathbb{Q}$-vector spaces

$$H^2(M_f,\mathbb{Q})_{alg} \longrightarrow \mathrm{Hom}(A_f^1,\ A_f^2)\otimes_{\mathbb{Z}}\mathbb{Q}.$$

Thus the problem of the determination of the Picard number of a Hilbert modular surface is reduced to the determination of $\mathrm{Hom}(A_f^1,\ A_f^2)$ for each primitive form f of weight 2.

In §5 of Chapter II, we recall the formalism constructing abelian varieties attached to Clifford algebras, following Satake [40], Kuga-Satake [20] and Deligne [10]. Applying this construction to the Hodge structure $H^2(M_f,\mathbb{Q})$, we have an abelian variety A(f) of dimension $4[K_f:\mathbb{Q}]$. Apply the results of §3, this abelian variety is found to be isogenous to a product $A_f^1\times A_f^1\times A_f^2\times A_f^2$ of two type of abelian varieties A_f^1 and A_f^2. In §6, we represent the period moduli of these abelian varieties A_f^1 and A_f^2 in terms of the periods of $H^2(M_f,\mathbb{Q})$. This explicit presentation of period moduli is used in the proof of Main Theorem B (§17.2). In §7, we prove Main Theorem A described above. In §8, we discuss the Hodge structure attached to symmetric Hilbert modular forms f, which we call selfconjugate forms in this paper. For such f, there exist always algebraic cycles in $H^2(M_f,\mathbb{Q})$, which implies that A_f^1 and A_f^2 are K_f-isogenous for such f. In §9, we discuss some results on non-selfconjugate forms f.

In Chapter III, we revise the Doi-Naganuma lifting of the real Nebentype cusp forms, and consider its adjoint mapping, applying the results of the previous paper [31]: In §10, we recall the Weil representation and the transformation formula of the theta series attached to quadratic forms. In §11, following the idea of Shintani [49], we construct real Nebentype elliptic modular cusp forms whose Fourier coefficients are period integrals of Hilbert modular cusp forms. §12 is a revision of Doi-Naganuma mapping via Zagier [55]. And in §13, we discuss the adjointness of this mapping and the the mapping defined in §11. This adjointness is interpreted as as a kind of period relation between the periods of real Nebentype elliptic modular forms and the periods of Hilbert modular forms, and used in the proof of Main Theorem B (cf. the proof §17.4 of Proposition 17.1).

In Chapter IV, we prove Main Theorem B, which claims that our method attaching abelian varieties A_f^i (i=1,2) to each primitive form f of weight 2, is functorial with respect to the Doi-Naganuma lifting. By using this theorem, we show that for symmetric primitive form f, the Hodge structure $H^2(M_f,\mathbb{Q})$ has no algebraic cycles more than constructed by Hirzebruch-Zagier (Theorem C in §18)

§14 is devoted to recall the period relation of Riemann for real Nebentype elliptic modular cusp forms. §15 is a preliminary to construct special kinds of 2-cycles of Hilbert modular surfaces, in order to show that the special values of L-functions of Hilbert modular cusp forms of weight 2 are represented by the period integrals along these special 2-cycles of the Hilbert modular surfaces (§16). In §17, we prove the following Main Theorem B.

Assume that the discriminant D of the real quadratic field F with class number 1 is a prime number. Let h be a real Nebentype elliptic modular cusp form of weight 2 with respect to $\Gamma_0(D)$ and with character $(\frac{D}{\ })$. Then by the theory of Shimura, for any common eigenform f of all Hecke operators, we can attach an abelian variety A_h of dimension $[K_h:\mathbb{Q}]$. Moreover A_h is isogenous to a product $B_h \times B_h$ of an abelian variety B_h (cf. Shimura [43], Chap.7).

<u>Main Theorem B</u>. <u>Let</u> f <u>be the a Hilbert modular cusp form of weight</u> 2 <u>with respect to</u> $SL_2(O_F)$, <u>which the Doi-Naganuma lifting of the real Nebentype primitive form</u> h <u>of weight</u> 2. <u>Then the abelian varieties</u> A_f^i (i=1,2) <u>are both isogenous to the abelian variety</u> B_h.

Applying this theorem and the results of Shimura [45], in §18 we show the following.

<u>Main Theorem C</u>. <u>Under the same assumption as that of Main Theorem</u> B, <u>for any symmetric primitive form</u> f <u>of weight</u> 2 <u>with respect to</u> $SL_2(O_F)$, <u>we have</u>

$$\mathrm{rank}_{\mathbb{Q}}\, H^2(M_f,\mathbb{Q})_{alg} = [K_f:\mathbb{Q}].$$

In §19, we discuss some examples on the ℓ-adic cohomology groups attached to primitive forms f of weight 2. By using the results of

Deligne [10] on the ℓ-adic cohomology groups of K3 surfaces, we shall show that the essential part of the second ℓ-adic cohomology group of a K3 type Hilbert modular surface (D=29, 37 or 41) is isomorphic to a tensor product of two copies of a direct factor of the first ℓ-adic cohomology group of certain modular curves, as Galois modules of $\mathrm{Gal}(\overline{\mathbb{Q}}/L)$ for a sufficiently large number field L.

§20 contains a remark on the Tate conjecture of Hilbert modular surfaces, and some conjectures and remarks on the field of definition of A_f^i (i=1,2), and on the relation of the ℓ-adic cohomology groups attached to f and A_f^i.

The main results of this paper are announced in [59].

Chapter I. Hodge structures attached to primitive forms of weight 2.

§0. Definitions and notations.

0.1. Let F be a real quadratic field with discriminant D. Throughout this paper we assume the class number of F is one. We regard F as a subfield of the real number field $\mathbb{R}$. For any element α of F, we denote by α' the conjugate of α over the rational number field $\mathbb{Q}$. Moreover we assume that there exists a unit ε_0 in the integer ring O_F of F with negative norm $N_{F/\mathbb{Q}}(\varepsilon_0)=\varepsilon_0\varepsilon_0'=-1$.

The embedding $F \longrightarrow \mathbb{R}\oplus\mathbb{R}$, obtained by a mapping $\alpha \longmapsto (\alpha,\alpha')$, induces an embedding $SL_2(F) \longrightarrow SL_2(\mathbb{R})\times SL_2(\mathbb{R})$ of the special linear group $SL_2(F)$ of degree 2 with entries in F into the product of the real special linear group $SL_2(\mathbb{R})$ of degree 2.

Let H be the complex upper half plane on which $SL_2(\mathbb{R})$ acts in the usual manner:

$$g(z)=\frac{az+b}{cz+d}, \quad \text{for } z\in H,\ g=\begin{pmatrix} a & b \\ c & d \end{pmatrix}\in SL_2(\mathbb{R}).$$

The product $SL_2(\mathbb{R})\times SL_2(\mathbb{R})$ acts on the product $H\times H$ factorwise. By composing this action with the embedding $SL_2(F) \longrightarrow SL_2(\mathbb{R})\times SL_2(\mathbb{R})$, we can define an action of $SL_2(F)$ on $H\times H$. Then the subgroup $SL_2(O_F)$ of $SL_2(F)$, which is a discrete subgroup of $SL_2(\mathbb{R})\times SL_2(\mathbb{R})$, acts properly discontinuously on $H\times H$. We denote by Γ the group $SL_2(O_F)/\{\pm 1\}$.

0.2. A holomorphic function $f(z)$ on $H\times H$ is called a Hilbert modular form of weight k, if it satisfies

$$f(g(z))=(\gamma z_1+\delta)^{-k}(\gamma' z_2+\delta')^{-k}f(z)$$

for any $g=\begin{pmatrix} \alpha & \beta \\ \gamma & \delta \end{pmatrix}\in SL_2(O_F)$ and $z=(z_1,z_2)\in H\times H$.

Recall that any Hilbert modular form has a Fourier expansion at each cusp of $SL_2(O_F)$. In our case, by the assumption that the class number of F is 1, $SL_2(O_F)$ has only one equivalence class of cusps. Therefore, $f(z)$ is called a cusp form, if and only if the constant term $a((0))$ of its Fourier expansion at infinity

$$\sum_{\nu\in O_{F+}\cup\{0\}} a((\nu))\exp[2\pi i(\nu\omega z_1+\nu'\omega' z_2)]$$

is zero. Here ω is a totally positive generator of the codifferent δ_F^{-1} of F, and O_{F+} is the set of the totally positive integers of F.

0.3. Let us recall the definition of Hecke operators. For any prime ideal $\mathfrak{p}=(\pi)$ of O_F with totally positive generator π, the action of the Hecke operator $T(\mathfrak{p})$ on the space of cusp forms $S_k(SL_2(O_F))$, is given by

$$f|T(\mathfrak{p})=N_{F/\mathbb{Q}}\mathfrak{p}^{k-1}\{f(\pi z_1,\pi' z_2)+N_{F/\mathbb{Q}}\pi^{-k}\sum_{\nu\in O_F/\mathfrak{p}} f(z_1+\frac{\nu}{\pi},z_2+\frac{\nu'}{\pi'})\}$$

for $f\in S_k(SL_2(O_F))$. More generally, for arbitrary integral ideal $\mathfrak{a}$ of O_F, we can define an action of $T(\mathfrak{a})$ similarly as the case of elliptic modular forms (cf. Shimura [43]).

0.4. We denote by $S_k(\Gamma_0(D),\varepsilon_D)$ the space of real Neben type elliptic modular cusp forms of weight k with respect to $\Gamma_0(D)=\{ \begin{pmatrix} a & b\\ c & d\end{pmatrix}\in SL_2(\mathbb{Z}) \mid c\equiv \bmod D \}$ and a character $\varepsilon_D(*)=(\frac{D}{*})$. This space will not appear in the first two chapters.

0.5. <u>Remark</u>. We define the Petersson metrics of $S_k(SL_2(O_F))$ and $S_k(\Gamma_0(D),\varepsilon_D)$ in a later section. Note then the normalization of these metrics is different from the conventional one.

0.6. The quotient analytic space $S=\Gamma\backslash(H\times H)$ is called a Hilbert modular surface, whose second cohomology group $H^2(S,\mathbb{Q})$ is the object of our investigation. The basic facts on this surface are found in Hirzebruch [16]. The Hecke operators $T(\mathfrak{a})$ are naturally regarded as algebraic correspondences of S.

0.7. An element f of $S_k(SL_2(O_F))$ is called a <u>primitive form</u>, if f is a common eigenform of all Hecke operators. Moreover f is called a <u>normalized primitive form</u>, if the fisrt Fourier coefficient $a((1))=a(O_F)$ of the primitive form f is 1.

§1. Hodge structures attached to Hilbert modular surfaces.

In this section we recall basic facts on the Hodge structures of the rational second cohomology groups of Hilbert modular surfaces S.

1.0. Let us summarize the theory of the mixed Hodge structures of Deligne [11] briefly. Let X be a separated scheme of finite type over $\mathbb{C}$. Then the cohomology groups $H^i(X,\mathbb{Q})$ of X are given two filtrations: an increasing filtration $\{W_k\}_{k\in\mathbb{Z}}$ on $H^i(X,\mathbb{Q})$ called weight filtration, and decreasing filtration $\{F^p\}_{p\in\mathbb{Z}}$ on $H^i(X,\mathbb{C})=H^i(X,\mathbb{Q})\otimes_{\mathbb{Q}}\mathbb{C}$ called Hodge filtration. And these filtrations satisfy the followings:

(i) $W_j(H^i(X,\mathbb{Q}))=0$ for $j<0$, and $W_j(H^i(X,\mathbb{Q}))=H^i(X,\mathbb{Q})$ for $j\geqq 2i$.

(ii) The induced filtration of $\{F^p\}_{p\in\mathbb{Z}}$ on

$$Gr_j^W H^i(X,\mathbb{C})=W_jH^i(X,\mathbb{C})/W_{j+1}H^i(X,\mathbb{C})$$

defines a Hodge filtration of a homogeneous Hodge structure of weight j. Namely $F^p(Gr_j^W H^i(X,\mathbb{C}))\oplus \overline{F}^{j-p+1}(Gr_j^W H^i(X,\mathbb{C}))\cong Gr_j^W H^i(X,\mathbb{C})$ for any p, and

$$Gr_j^W H^i(X,\mathbb{C}) = \bigoplus_{p=0}^{j} F^p\cap \overline{F}^{j-p}.$$

It is known that for any smooth scheme X,

$$W_jH^i(X,\mathbb{Q})=0 \qquad \text{for } j<i,$$

and that for any proper scheme X,

$$W_jH^i(X,\mathbb{Q})=H^i(X,\mathbb{Q}) \qquad \text{for } j\geqq i.$$

Suppose that X is smooth, and $\tilde{X}$ is a smooth compactification of X, then

$$W_iH^i(X,\mathbb{Q})=\text{Image}(H^i(\tilde{X},\mathbb{Q}) \longrightarrow H^i(X,\mathbb{Q})).$$

When X is a <u>rational</u> <u>homology</u> <u>manifold</u>, similar result holds.

1.1. The quotient space $S=\Gamma\backslash(H\times H)$ has a natural structure of complex analytic space of dimension 2. By Baily-Borel [4], or by Rapoprt [36], it is known that S has a natural structure of a quasi-projective algebraic surface. Corresponding to non-trivial torsion elements of Γ, S has finite number of quotient singularities.

Let us regard $H\times H$ as a subset of $P^1(\mathbb{C})\times P^1(\mathbb{C})$, where $P^1(\mathbb{C})=\mathbb{C}\cup\{\infty\}$. Then the rational boundary components of Γ is given by a subset $\{(\alpha,\alpha') \mid \alpha\in F\}\cup\{(\infty,\infty)\}$ of $P^1(\mathbb{C})\times P^1(\mathbb{C})$. Let $\overline{S}$ be the standard compactification of S, obtained by attaching the finite number of points to S, which are corresponding to Γ-equivalence classes of cusps $P^1(F)$. The algebraic surface $\overline{S}$ has singularities along these points corresponding to cusps. Resolutions of these singularities are investigated by Hirzebruch [16].

1.2. We denote by $C=\{c_1, c_2, \ldots, c_h\}$ the set of all singular points corresponding to the cusps on $\overline{S}$. Let $\tilde{S} \xrightarrow{p} \overline{S}$ be a resolution of all singularities along cusps. By definition, the restriction of p to $S\text{-}p^{-1}(C)$ induces an isomorphism

$$\tilde{S}-p^{-1}(C) \cong \overline{S}-C.$$

$\tilde{S}$ has no singularities except quotient singularities corresponding to the torsion elements of Γ. Since $\tilde{S}$ has only rational singularities, $\tilde{S}$ is a compact rational homology manifold. Therefore we have a perfect pairing

$$H_2(\tilde{S},\mathbb{Q}) \times H_2(\tilde{S},\mathbb{Q}) \longrightarrow \mathbb{Q}$$

via Poincaré duality.

Take as $\tilde{S}$ the resolution constructed by Hirzebruch [16]. For each c_i $(1 \leqq i \leqq h)$, we denote by L_i the subspace of $H_2(S,\mathbb{Q})$ generated by the fundamental classes of the algebraic curves, obtained as irreducible components of $p^{-1}(c_i)$. Then, as shown in 2.4 (p.212) of [16], the restriction of the intersection form $\langle\ ,\ \rangle$ to each L_i defines a non-degenerate bilinear form on L_i. Clearly $\langle L_i, L_j\rangle=0$, if $i \neq j$. Hence the restriction of $\langle\ ,\ \rangle$ to $L=\bigoplus_{i=1}^{h} L_i$ defines a non-degenerate bilinear form on L. Since $\tilde{S}$ is smooth in neighbourhoods of the generators of L, for any simplex $\gamma:\Delta^2 \longrightarrow \tilde{S}$ satisfying $\langle\gamma,\ell\rangle=0$ for all $\ell \in L$, we can find a simplex $\gamma':\Delta^2 \longrightarrow S \subset \tilde{S}$ homologous to γ. Here Δ^2 stands for the standard 2-simplex.

Therefore $H_2(\tilde{S},\mathbb{Q})$ is a direct sum of two subspaces L and

$$\mathrm{Image}(H_2(S,\mathbb{Q}) \longrightarrow H_2(\tilde{S},\mathbb{Q})),$$

which are mutually orthogonal with respect to the intersection form $\langle\ ,\ \rangle$. Especially, the restriction of $\langle\ ,\ \rangle$ to $\mathrm{Image}(H_2(S,\mathbb{Q}) \longrightarrow H_2(\tilde{S},\mathbb{Q}))$ is a non-degenerate bilinear form.

1.3. In this paragraph and the next, we shall see the weight filtration on the second cohomology group $H^2(S,\mathbb{Q})$ of S. In the first place, let us consider the cohomology groups of S, $\overline{S}$, and $\tilde{S}$. We denote by j the natural inclusion $j:S \hookrightarrow \tilde{S}$. Then the composition

$$p \circ j : S \longrightarrow \overline{S}$$

is also an inclusion mapping. We have a commutative diagram of cohomology groups

$$\begin{array}{ccccccc}
\longrightarrow H^2_c(S,\mathbb{Q}) & \xrightarrow{\ j_!\ } & H^2(\tilde{S},\mathbb{Q}) & \longrightarrow & H^2(\tilde{S}-S,\mathbb{Q}) & \longrightarrow \\
\wr\|\ \alpha & \circlearrowleft & \|\ \text{id.} & & & \\
H^2(\overline{S},\mathbb{Q}) & \xrightarrow{\ p^*\ } & H^2(\tilde{S},\mathbb{Q}) & & & \\
 & {\searrow}^{j^*p^*} \quad {\swarrow}_{j^*} & & & & \\
 & H^2(S,\mathbb{Q}) & & & &
\end{array}$$

Here the top horizontal sequence is a long exact sequence of cohomology groups with compact supports and the isomorphism α is induced from a long exact sequence

$$\longrightarrow \underset{\substack{\| \\ 0}}{H^1(\overline{S}-S,\mathbb{Q})} \longrightarrow H^2_c(S,\mathbb{Q}) \xrightarrow{\ \alpha\ } H^2(\overline{S},\mathbb{Q}) \longrightarrow \underset{\substack{\| \\ 0}}{H^2(\overline{S}-S,\mathbb{Q})} \longrightarrow$$

Let L^* be the subspace of $H^2(\tilde{S},\mathbb{Q})$ generated by the Poincaré duals of the algebraic cycles corresponding to the irreducible components of $p^{-1}(c_i)$ $(1\leqq i\leqq h)$. Then,

$$H^2(S,\mathbb{Q})=p^*H^2(\overline{S},\mathbb{Q}) + L^*,$$

and $j^*(L^*)=0$, for $j^*:H^2(\tilde{S},\mathbb{Q}) \longrightarrow H^2(S,\mathbb{Q})$.
Therefore $j^*p^*H^2(\overline{S},\mathbb{Q})=j^*H^2(\tilde{S},\mathbb{Q})=j^*j_!H^2_c(S,\mathbb{Q})$. This common image of j^*p^*, j^*, and $j^*j_!$ coincides with $W_2H^2(S,\mathbb{Q})$.

Since S is a rational homology manifold, $W_1H^2(S,\mathbb{Q})=0$, by Theorem (8.2.4) (iv) of Deligne [11, III].

Noting that the inverse image $p^{-1}(c_i)$ of each cusp singularity c_i $(1\leqq i\leqq h)$ by means of the resolution of Hirzebruch [16], is a non-smooth stable curve of genus 1, we can find the weight filtration of the mixed Hodge structure of $H^2(S,\mathbb{Q})$ as follows:

$$W_kH^2(S,\mathbb{Q})=0 \quad \text{for } k\leqq 1,$$
$$W_2H^2(S,\mathbb{Q})=W_3H^2(S,\mathbb{Q}),$$
$$W_kH^2(S,\mathbb{Q})=H^2(S,\mathbb{Q}) \quad \text{for } k\geqq 4,$$

and

$$Gr^W_4H^2(S,\mathbb{Q})=H^2(S,\mathbb{Q})/W_2H^2(S,\mathbb{Q})=\overset{h}{\oplus}\ \mathbb{Q}(-2).$$

By Poincaré duality

$$H^2(S,\mathbb{Q}) \times H^2_c(S,\mathbb{Q}) \longrightarrow \mathbb{Q}(-2),$$

the weight filtration of $H^2_c(S,\mathbb{Q})=H^2(\overline{S},\mathbb{Q})$ is as follows:

$$W_kH^2(\overline{S},\mathbb{Q})=0 \quad \text{for } k< 0,$$
$$W_1H^2(\overline{S},\mathbb{Q})=W_0H^2(\overline{S},\mathbb{Q}),$$
$$W_kH^2(\overline{S},\mathbb{Q})=H^2(\overline{S},\mathbb{Q}) \quad \text{for } k\geqq 2,$$
$$Gr^W_0H^2(\overline{S},\mathbb{Q})=W_0H^2(\overline{S},\mathbb{Q})=\overset{h}{\oplus}\ \mathbb{Q}(0),$$

and

$$Gr_2^W H^2(\overline{S},\mathbb{Q})=W_2H^2(\overline{S},\mathbb{Q})/W_1H^2(\overline{S},\mathbb{Q})$$

has a homogeneous Hodge structure of weight 2, dual to $W_2H^2(S,\mathbb{Q})(2)$.

1.4. In a later paragraph, we need to know the kernel of

$$j^*p^*:H^2(\overline{S},\mathbb{Q}) \longrightarrow H^2(S,\mathbb{Q}).$$

We can descibe this kernel in terms of the cohomology group of the "boundary" ∂S, which is introduced by Harder [14].

Let λ be a cusp of Γ, which is written as $\lambda=\rho/\sigma$ with mutually coprime integers ρ and σ of O_F. Following Siegel [50, Chap. III. §2], for $\lambda=\rho/\sigma\neq\infty$, and $z=(z_1, z_2)=(x_1+\sqrt{-1}y_1, x_2+\sqrt{-1}y_2)\in H\times H$, we put

$$\ell(z,\lambda)=y_1^2y_2^2\{(\sigma x_1-\rho)^2+\sigma^2y_1^2\}^{-1}\{(\sigma' x_2-\rho')^2+\sigma'^2y_2^2\}^{-1}$$

and for $\lambda=\infty$, we put

$$\ell(z,\infty)=y_1y_2.$$

Then $\ell(z,\lambda)$ has an invariance property

$$\ell(\gamma(z),\gamma(\lambda))=\ell(z,\lambda) \quad \text{for any } \gamma\in\Gamma.$$

It is known that for sufficiently large real numbers M,

$$\ell(z,\lambda)=\ell(z,\mu)=M$$

implies $\lambda=\mu$.

Put

$$S_M=\Gamma\backslash\{ z\in H\times H \mid \ell(z,\lambda)\leqq M \text{ for all cusps } \lambda \},$$

and

$$\partial S_M=\Gamma\backslash\{ z\in H\times H \mid \ell(z,\lambda)=M \text{ for some cusp } \lambda \}.$$

For sufficiently large M, we have an isomorphism

$$\partial S_M=\Gamma_\infty\backslash\{z\in H\times H \mid \ell(z,\infty)=M \},$$

where

$$\Gamma_\infty=\{ \gamma\in\Gamma \mid \gamma(\infty)=\infty \}.$$

For sufficiently large M, S_M is a homotopy equivalence of S, and by excision theorem,

$$H^2(\overline{S},\mathbb{Q})=H^2(\overline{S} \bmod C,\mathbb{Q})=H^2(S_M \bmod \partial S_M,\mathbb{Q}).$$

Therefore we have an exact sequence

$$\begin{array}{ccccccc}
\longrightarrow H^2(S_M \bmod \partial S_M,\mathbb{C}) & \longrightarrow & H^2(S_M,\mathbb{C}) & \xrightarrow{r} & H(\partial S_M,\mathbb{C}) & \longrightarrow \\
\wr\| & \circlearrowleft & \wr\| & & & \\
H^2(\overline{S},\mathbb{C}) & \longrightarrow & H^2(S,\mathbb{C}) & & &
\end{array}$$

1.5. In this and subsequent several paragraphs, we shall investigate the Hodge decomposition of $W_2H^2(S,\mathbb{Q})\otimes_{\mathbb{Q}}\mathbb{C}$. First of all, let us see the

(2,0)-type component of $W_2H^2(S,\mathbb{C})$.

Suppose that $f(z_1,z_2)$ is a Hilbert modular cusp form of weight 2, and consider a Γ-invariant 2-form ω_f on $H\times H$ defined by

$$\omega_f=(2\pi i)^2 f(z_1,z_2)dz_1\wedge dz_2.$$

Let us define the de Rham cohomology group $H^2_{DR}(S,\mathbb{C})$ of S, by

$$H^2_{DR}(S,\mathbb{C}) \underset{dfn}{=} H^2_{DR}(S',\mathbb{C})^{\Gamma/\Gamma'},$$

where Γ' is a torsion-free normal subgroup of Γ with finite index, and $S'=\Gamma'\backslash(H\times H)$.
Clearly $H^2_{DR}(S,\mathbb{C})$ does not depend on the choice of Γ'. The 2-form ω_f defines an element of $H^2_{DR}(S,\mathbb{C})$.

Put $S_e=S-\{\text{quotient singularities}\}$, and let $\tilde{S}_e$ be a smooth compactification of S_e, obtained by resolving these quotient singularities on $\tilde{S}$. Then we have a diagram

$$\begin{array}{ccc} S & \xrightarrow{\ j\ } & \tilde{S} \\ i\uparrow & \circlearrowleft & \uparrow q \\ S_e & \xrightarrow[j_e]{} & \tilde{S}_e \end{array}$$

where i, j, and j_e are open immersion, and q is proper surjective. Then it is shown by Lemma in Section 3.6 (p.233) of [16], that we can prolong ω_f to a holomorphic 2-form on $\tilde{S}_e$, and by this correspondence the space of holomorphic 2-forms on $\tilde{S}_e$ is canonically isomorphic to the space $S_2(SL_2(O_F))$ of Hilbert modular cusp forms of weight 2. Especially $\dim_{\mathbb{C}} S_2(SL_2(O_F))$ is equal to the geometric genus p_g of $\tilde{S}_e$, which is a birational invariant.

Consider the commutative diagram of homogeneous Hodge structures of weight 2

$$\begin{array}{ccc} H^2(\tilde{S},\mathbb{Q}) & \xrightarrow{\ j^*\ } & W_2H^2(S,\mathbb{Q}) \\ q^*\downarrow & \circlearrowleft & \downarrow i^* \\ H^2(\tilde{S}_e,\mathbb{Q}) & \xrightarrow[j_e^*]{} & W_2H^2(S_e,\mathbb{Q}). \end{array}$$

Then it is easy to see that the kernels of epimorphisms j^*, j_e^* and the cokernels of monomorphisms i^*, q^* are (1,1) type Hodge structures generated by algebraic cycles. Therefore $\omega_f\in H^2_{DR}(S,\mathbb{C})$ defines an element of (2,0) type in $W_2H^2(S,\mathbb{C})$, and we have an identification

$$F^2W_2H^2(S,\mathbb{C})=\{\ \omega_f \mid f\in S_2(SL_2(O_F))\},$$

where $F^{\cdot}$ is the Hodge filtration.

By Hodge symmetry,

$$\overline{F}^2 W_2 H^2(S,C) = \{\ \overline{\omega}_f \ \mid f \in S_2(SL_2(O_F))\ \}.$$

1.6. Let us find out (1,1) type component of the Hodge structure $W_2H^2(S,\mathbb{C})$.

Fix a unit ε_0 of O_F with $\varepsilon_0>0$, $\varepsilon_0'<0$. Define elements η_1, η_2, and $\eta_{f,1}$, $\eta_{f,2}$ of $H^2_{DR}(S,C)$ for $f\in S_2(SL_2(O_F))$, by

$$\eta_1=\frac{1}{2\pi i}\,\frac{dz_1\wedge\overline{dz_1}}{y_1^2},\qquad \eta_2=\frac{1}{2\pi i}\,\frac{dz_2\wedge\overline{dz_2}}{y_2^2}\qquad (z_i=x_i+\sqrt{-1}y_i,\ i=1,2),$$

and

$$\eta_{f,1}=(2\pi i)^2 f(\varepsilon_0 z_1,\varepsilon_0'\overline{z}_2)dz_1\wedge\overline{dz}_2,\qquad \eta_{f,2}=(2\pi i)^2 f(\varepsilon_0'\overline{z}_1,\varepsilon_0 z_2)\overline{dz}_1\wedge dz_2,$$

for $f\in S_2(SL_2(O_F))$.

We can readily check that these forms in $H^2(S,\mathbb{C})=H^2_{DR}(S,\mathbb{C})$ belong to the kernel of

$$H^2(S,\mathbb{C})\xrightarrow{\ r\ } H^2(\partial S_M,\mathbb{C}).$$

In fact, for a generator γ_0 of $H^2(\partial S_M,\mathbb{Z})\cong\mathbb{Z}$, and $\phi\in H^2_{DR}(S,\mathbb{C})$, we have

$$<\gamma_0,\ r(\phi)>=\int_{\substack{y_1=y_2=\text{constant}\\ (x_1,x_2)\in\mathbb{R}^2 \bmod O_F}} \phi(z_1,z_2),$$

which is "the constant term" of the Fourier expansion of ϕ. Evidently this integral is zero for $\phi=\eta_1$, η_2, $\eta_{f,1}$, and $\eta_{f,2}$. Thus η_1, η_2, $\eta_{f,1}$, and $\eta_{f,2}$ belong to

$$\text{Image}(H^2(\overline{S},\mathbb{C})\longrightarrow H^2(S,\mathbb{C}))=W_2H^2(S,\mathbb{C}).$$

1.7. <u>Proposition</u>. $\dim_{\mathbb{Q}} W_2H^2(S,\mathbb{Q})=4p_g+2$.

<u>Proof</u>. It is known that the first Betti number of S $b_1(S)=0$ (cf.[16], for example). And a long exact sequence of relative cohomology

$$\begin{array}{ccccccccc} 0 \longrightarrow & H^0(S,\mathbb{Q}) & \longrightarrow & H^0(\partial S_M,\mathbb{Q}) & \longrightarrow & H^1_c(S,\mathbb{Q}) & \longrightarrow & H^1(S,\mathbb{Q}) & \longrightarrow \\ & \wr\Vert & & \wr\Vert & & & & \wr\Vert & \\ & \mathbb{Q} & & \mathbb{Q} & & & & \{0\} & \end{array}$$

implies $H^1_c(S,\mathbb{Q})=0$, accordingly $b_3(S)=0$ by Poincaré duality. Therefore the Euler characteristic $e(S)$ of S is $b_0(S)+b_2(S)=1+b_2(S)$. Since $b_2(S)=\dim_{\mathbb{Q}} W_2H^2(S,\mathbb{Q})+\dim_{\mathbb{Q}} Gr_4^W H^2(S,\mathbb{Q})$, and $\dim_{\mathbb{Q}} Gr_4^W H^2(S,\mathbb{Q})$ is equal to the number of cusps, we have

$$e(S)=\dim_{\mathbb{Q}} W_2H^2(S,\mathbb{Q})+2.$$

On the other hand, by the formula (43) of Theorem 3.9 of [16],

$$e(S)=4(1-q+p_g),$$

where q (resp. p_g) is the irregularity (resp. the geometric genus) of the smooth compact model of S.

Since Hilbert modular surfaces are regular,

$$e(S)=4+4p_g.$$

Hence,

$$\dim_{\mathbb{Q}} W_2H^2(S,\mathbb{Q})=4p_g+2. \qquad \text{q.e.d.}$$

1.8. Since

$$\int_S \eta_1\wedge\eta_1=0, \qquad \int_S \eta_2\wedge\eta_2=0, \quad \text{and} \qquad \int_S \eta_1\wedge\eta_2=\text{volume of } S\neq 0,$$

by de Rham's theorem,

$$\langle\eta_1,\eta_1\rangle=\langle\eta_2,\eta_2\rangle=0, \quad \text{and} \quad \langle\eta_1,\eta_2\rangle\neq 0,$$

where $\langle\ ,\ \rangle$ is the intersection form on $W_2H^2(S,\mathbb{C})$.

Similarly, we have

$$\langle\omega_f,\omega_g\rangle=\langle\omega_f,\eta_{g,1}\rangle=\langle\omega_f,\eta_{g,2}\rangle=\langle\eta_{f,1},\eta_{g,1}\rangle=\langle\eta_{f,1},\overline{\omega}_g\rangle=\langle\eta_{f,2},\overline{\omega}_g\rangle$$
$$=\langle\eta_{f,2},\eta_{g,2}\rangle=\langle\overline{\omega}_f,\overline{\omega}_g\rangle=0,$$

for any elements f, g of $S_2(SL_2(O_F))$, and

$$\langle\omega_f,\eta_1\rangle=\langle\omega_f,\eta_2\rangle=\langle\eta_{f,1},\eta_1\rangle=\langle\eta_{f,2},\eta_1\rangle=\langle\eta_{f,1},\eta_2\rangle=\langle\eta_{f,2},\eta_2\rangle$$
$$=\langle\overline{\omega}_f,\eta_1\rangle=\langle\overline{\omega}_f,\eta_2\rangle=0,$$

for any element f of $S_2(SL_2(O_F))$.

As shown in Section 1.2, the intersection form is non-degenerate over the dual space $\mathrm{Image}(H_2(S,\mathbb{Q}) \longrightarrow H_2(\tilde{S},\mathbb{Q}))$ of $W_2H^2(S,\mathbb{Q})$. Therefore, $W_2H^2(S,\mathbb{Q})\otimes_{\mathbb{Q}}\mathbb{C}$ is a direct sum of subspaces

$$\{\ \omega_f \mid f\in S_2(SL_2(O_F))\ \},$$
$$\{\ \eta_{f,1} \mid f\in S_2(SL_2(O_F))\ \}\oplus\{\ \eta_{f,2} \mid f\in S_2(SL_2(O_F))\ \}\oplus \mathbb{C}\eta_1\oplus\mathbb{C}\eta_2,$$

and $\{\ \overline{\omega}_f \mid f\in S_2(SL_2(O_F))\ \}$,

which gives the Hodge decomposition of $W_2H^2(S,\mathbb{C})$.

1.9. <u>Proposition</u>. <u>The elements</u> η_1 <u>and</u> η_2 <u>of</u> $W_2H^2(S,\mathbb{C})$ <u>represent elements of</u> $W_2H^2(S,\mathbb{Q})$. <u>Or equivalently</u>,

$$\int_\gamma \eta_i\in\mathbb{Q}, \quad \underline{\text{for any}} \quad \gamma\in H_2(S,\mathbb{Q}).$$

<u>Proof</u>. Let L_1, L_2 be the line bundles on S, corresponding to the automorphy factors given by

$$j_1(g,z)=(\gamma z_1+\delta)^2, \quad j_2(g,z)=(\gamma' z_2+\delta')^2$$

for $g=\begin{pmatrix}\alpha & \beta\\ \gamma & \delta\end{pmatrix}\in SL_2(O_F)$ and $z=(z_1,z_2)$, respectively.
Then these line bundles are extendable to line bundles $\tilde{L}_i$ on a toroidal desingularization $\tilde{S}$ of $\overline{S}$ (cf. van der Geer and Ueno [52]).
By definition, the pull-back L_i^* of L_i to H×H is trivialized via $j_i(g,z)$. Put $g(z)=(w_1,w_2)$. Then,

$$\mathrm{Im}(w_i)=|j_i(g,z)|^{-2}\mathrm{Im}(z_i) \qquad (i=1,2),$$

where Im(x) stands for the imaginary part of a complex number x.
Therefore we can define a Hermitian metric h_i on L_i by

$$h_i(s_i(z),s_i(z))=(\mathrm{Im}(z_i))^{-2}|s_i(z)|^2 \quad (i=1,2),$$

where $s_i:U \longrightarrow L_i^* \cong \mathbb{C}$ is a Γ-invariant section on an open subset U of H×H.

We can check that these metrics are "good" in the sense of Mumford [28] (see Definition of p.242). The Chern form $c_1(L_i,h_i)$ is given by $2\eta_i$ (i=1,2). By Theorem 1.4 of [28], the current $[\eta_i]$ given by

$$\lim_{M\to\infty}\int_{S_M}\eta_i\wedge\psi,$$

for a C^∞ 2-form ψ on $\tilde{S}$, defines the same element as $c_1(\tilde{L}_i)$ in $H^2(\tilde{S},\mathbb{C})$.
Since $c_1(\tilde{L}_i)\in H^2(\tilde{S},\mathbb{Z})$,

$$\int_\gamma \eta_i=\langle j_*(\gamma),c_1(\tilde{L}_i)\rangle\in\mathbb{Q},$$

where $j_*:H_2(S,\mathbb{Q}) \longrightarrow H_2(\tilde{S},\mathbb{Q})$. q.e.d.

1.10. Define $\tilde{H}_2(S,\mathbb{C})$ by

$$\tilde{H}_2(S,\mathbb{Q})=\mathrm{Coimage}(H_2(S,\mathbb{Q}) \longrightarrow H_2(\tilde{S},\mathbb{Q})),$$

and equip this space with an intersection form < , > by means of the canonical isomorphism

$$\tilde{H}_2(S,\mathbb{Q})\cong\mathrm{Image}(H_2(S,\mathbb{Q}) \longrightarrow H_2(\tilde{S},\mathbb{Q}))\hookrightarrow H_2(\tilde{S},\mathbb{Q}).$$

By a standard arguement using the results of resolution of singularities, we can show that this intersection form does not depend upon the choice of S. The space $W_2H^2(S,\mathbb{C})$ is canonically dual to $\tilde{H}_2(S,\mathbb{Q})$.

<u>Definition</u>. $$H_2^{sp}(S,\mathbb{Q}) \underset{\mathrm{dfn}}{=} \{\ \gamma\in\tilde{H}_2(S,\mathbb{Q})\ \mid\ \int_\gamma\eta_1 = \int_\gamma\eta_2 =0\ \},$$

$$H^2_{sp}(S,\mathbb{Q}) \underset{\mathrm{dfn}}{=} \{\ \omega\in W_2H^2(S,\mathbb{Q})\ \mid\ \langle\eta_1,\omega\rangle=0,\quad \langle\eta_2,\omega\rangle=0\ \}.$$

Clearly $H^2_{sp}(S,\mathbb{Q})$ is a polarized sub-Hodge structure of $W_2H^2(S,\mathbb{Q})$, such that

$$W_2H^2(S,\mathbb{Q})=H^2_{sp}(S,\mathbb{Q}) \oplus \mathbb{Q}(-1) \oplus \mathbb{Q}(-1),$$

where $\mathbb{Q}(-1)$ is the Hodge structure of Tate of weight 2.

1.11. Let us consider the action of the Hecke operators on the homology and cohomology groups of S. We denote by H the ring of Hecke operators.

Let α be an element of the integral matrices $M_2(O_F)$ of size 2 over O_F with $\det(\alpha)$ totally positive. Then we consider an algebraic correspondence $T_{\Gamma\alpha\Gamma}$ of S for each double coset $\Gamma\alpha\Gamma$, defined by

$$T_{\Gamma\alpha\Gamma}=\{\ (z,\ \alpha(z))\in(H\times H)\times(H\times H) \bmod \Gamma\times\Gamma \ \mid\ z\in H\times H\ \}\subset S\times S.$$

Let pr_1 and pr_2 be the first and the second projections of $T_{\Gamma\alpha\Gamma}$ to S. Then both projections are finite and flat. And moreover both are étale except over a finite number of points of S. Therefore we can define the trace morphisms

$$pr_{1*} \text{ and } pr_{2*}:H^2(T_{\Gamma\alpha\Gamma},\mathbb{Q}) \longrightarrow H^2(S,\mathbb{Q})$$

induced from

$$pr_1^* \text{ and } pr_2^* :H^2_c(S,\mathbb{Q}) \longrightarrow H^2_c(T_{\Gamma\alpha\Gamma},\mathbb{Q})$$

by Poincaré duality, respectively.

For $\gamma\in H_2(S,\mathbb{Q})$ we define $(T_{\Gamma\alpha\Gamma})_*(\gamma)$ by

$$(T_{\Gamma\alpha\Gamma})_*(\gamma)=pr_2(pr_1^{-1}(\gamma)),$$

and for $\omega\in H^2_{DR}(S,\mathbb{C})$ we define $(T_{\Gamma\alpha\Gamma})^*(\omega)$ by

$$(T_{\Gamma\alpha\Gamma})^*(\omega)=pr_{1*}pr_2^*(\omega).$$

It is easy to check that we have

$$\int_{(T_{\Gamma\alpha\Gamma})_*(\gamma)} \omega = \int_{\gamma} (T_{\Gamma\alpha\Gamma})^*(\omega),$$

for any $\gamma\in H_2(S,\mathbb{Q})$ and $\omega\in H^2_{DR}(S,\mathbb{C})$.

Let $\deg:H \longrightarrow \mathbb{Z}$ be the degree homomorphism of the Hecke algebra H. Then, we can easily check that H acts on $H_2(\partial S_M,\mathbb{Q})$ by

$$(T_{\Gamma\alpha\Gamma})_*(\gamma_0)=\deg(\Gamma\alpha\Gamma)\gamma_0,$$

for any $\gamma_0\in H_2(\partial S_M,\mathbb{Q})$. Therefore H acts on $\tilde{H}_2(S,\mathbb{Q})$ on passing to the quotient. Since $T_{\mathfrak{a}}$ is a symmetric correspondence for any ideal $\mathfrak{a}$ of O_F, we have

$$\langle\gamma_1,(T_{\mathfrak{a}})_*(\gamma_2)\rangle=\langle(T_{\mathfrak{a}})_*(\gamma_1),\gamma_2\rangle,$$

for any $\gamma_1, \gamma_2 \in \tilde{H}_2(S,\mathbb{Q})$.

Since T are algebraic correspondences, the action of $\mathcal{H}$ on $W_2H^2(S,\mathbb{Q})$ holds its Hodge structure, and for η_1 and η_2, $\mathcal{H}$ acts by

$$(T_{\Gamma\alpha\Gamma})^*(\eta_i)=\deg(\Gamma\alpha\Gamma)\eta_i \qquad (i=1,2).$$

Thus $\mathcal{H}$ acts also on $H^2_{sp}(S,\mathbb{Q})$ and $H_2^{sp}(S,\mathbb{Q})$ by restriction.

Summing up all these results, we have the following main theorem of this section.

1.12. Theorem. Define the subspaces $H^{2,0}$, $H^{1,1}$, and $H^{0,2}$ of $H^2_{DR}(S,\mathbb{C})$ by

$$H^{2,0}=\{\ \omega_f \mid f\in S_2(SL_2(O_F))\ \},$$
$$H^{1,1}=\{\ \eta_{f,1} \mid f\in S_2(SL_2(O_F))\ \}\oplus\{\ \eta_{f,2} \mid f\in S_2(SL_2(O_F)\ \},$$
$$H^{0,2}=\{\ \bar{\omega}_f \mid f\in S_2(SL_2(O_F)\ \}.$$

Then the Hecke algebra $\mathcal{H}$ acts on the homogeneous polarized Hodge structure $W_2H^2(S,\mathbb{Q})$ of weight 2 as endomorphisms of Hodge structure. And moreover we have decompositions of $\mathcal{H}$-modules

$$W_2H^2(S,\mathbb{Q})=H^2_{sp}(S,\mathbb{Q})\oplus\mathbb{Q}(-1)\oplus\mathbb{Q}(-1),$$

and

$$H^2_{sp}(S,\mathbb{Q})\otimes_{\mathbb{Q}}\mathbb{C}=H^{2,0}\oplus H^{1,1}\oplus H^{0,2}.$$

Remark. As a matter of fact, the above result is already proved by Harder [14], by using the square integrale cohomology classes (more detailed exposition is found in Hida [15]). But for geometric application, we need to make sure its relation with the Hodge structure of a smooth proper model of S.

1.13. Remark. Evidently from construction, we have an isomorphism of homogeneous rational polarized Hodge structures

$$H^2_{sp}(S,\mathbb{Q})\oplus\{\overset{a}{\oplus}\,\mathbb{Q}(-1)\} = H^2(S^*,\mathbb{Q})\oplus\{\overset{b}{\oplus}\,\mathbb{Q}(-1)\},$$

for any smooth proper model S^* of S. Here a, b are some natural integers.

§2. Hodge structures attached to primitive forms of weight 2.

In the previous section, we constructed a homogeneous polarized Hodge structure $H^2_{sp}(S,\mathbb{Q})$ of weight 2, attached to the Hilbert modular surface S. In this section, we decompose this Hodge structure into eigenspaces of the Hecke algebra H. If we denote by K_f the field of eigenvalues of f for any primitive Hilbert modular cusp form f of weight 2, then we can attach to such f a rational polarized Hodge structure $H^2(M_f,\mathbb{Q})$ of weight 2 and of rank $4[K_f:\mathbb{Q}]$ over $\mathbb{Q}$, on which K_f acts as endomorphisms of Hodge structure.

2.1. We write H_0 for a subring of the endomorphism ring $\mathrm{End}(H^2_{sp}(S,\mathbb{Q}))$ of the Hodge structure $H^2_{sp}(S,\mathbb{Q})$, generated by the identity and the images of the elements of H over $\mathbb{Q}$. Since H is commutative, H_0 is a commutative ring with unity, which is of finite dimension over $\mathbb{Q}$.

In order to show that $\mathbb{Q}$-algebra H_0 is semisimple, it suffices to show that $H_0\otimes_{\mathbb{Q}}\mathbb{C}$ is semisimple. Since the elements of H_0 hold Hodge type, by means of the Hodge decomposition of $H^2_{sp}(S,\mathbb{Q})\otimes_{\mathbb{Q}}\mathbb{C}$, we have

$$H_0\otimes_{\mathbb{Q}}\mathbb{C} \hookrightarrow \mathrm{End}_{\mathbb{C}}(H^2_{sp}(S,\mathbb{C})) \subset \mathrm{End}(H^{2,0}) \oplus \mathrm{End}(H^{1,1}) \oplus \mathrm{End}(H^{0,2}).$$

Let us consider the composition

$$H_0\otimes_{\mathbb{Q}}\mathbb{C} \hookrightarrow \mathrm{End}_{\mathbb{C}}(H^2_{sp}(S,\mathbb{C})) \xrightarrow{\text{projection}} \mathrm{End}(H^{2,0}),$$

which is injective, because by Theorem 1.12 the $H_0\otimes_{\mathbb{Q}}\mathbb{C}$-module $H^2_{sp}(S,\mathbb{C})$ is a direct sum of four $H_0\otimes_{\mathbb{Q}}\mathbb{C}$-modules isotypic to the (2,0) component $H^{2,0}$.

Since $S_2(SL_2(O_F))$ has a basis consisting of eigenforms of all Hecke operators, $H_0\otimes_{\mathbb{Q}}\mathbb{C}$ is a direct sum of the complex number field $\mathbb{C}$. Moreover the following theorem implies that

$$H_0\otimes_{\mathbb{Q}}\mathbb{C} \cong \overset{p_g}{\oplus}\ \mathbb{C},$$

where $p_g=\dim_{\mathbb{C}} S_2(SL_2(O_F))$.

2.2. <u>Multiplicity One Theorem.</u> <u>Let f_1 and f_2 be two primitive forms in $S_2(SL_2(O_F))$. Suppose that f_1 and f_2 have the same eigenvalues for all Hecke operators. Then f_1 is a constant multiple of f_2.</u>

<u>Proof.</u> Well known. In fact, a stronger result (strong multiplicity one theorem) is found in Miyake [24], for example. q.e.d.

This theorem implies that $H_0\otimes_{\mathbb{Q}}\mathbb{C}$ is a commutative semisimple algebra of rank p_g over $\mathbb{C}$, and that $S_2(SL_2(O_F)$ is a free $H_0\otimes_{\mathbb{Q}}\mathbb{C}$-module of rank 1.

Hence, H_0 is a commutative semisimple $\mathbb{Q}$-algebra of rank p_g, accordingly a direct sum of fields. Since $H^2_{sp}(S,\mathbb{C})$ is a free $H_0 \otimes_{\mathbb{Q}} \mathbb{C}$-module of rank 4. Therefore $H^2_{sp}(S,\mathbb{Q})$ is a free module of rank 4 over the semilocal ring H_0.

Let

$$H_0 = \bigoplus_{i=1}^{m} K_i$$

be a decomposition of H_0 as a direct sum of algebraic number fields K_i $(1 \leqq i \leqq m)$.

Put $d_i = [K_i : \mathbb{Q}]$. Then $\sum_{i=1}^{m} d_i = p_g$.

Let $\{ e_1, \ldots, e_m \}$ be the system of primitive idempotents in H_0, corresponding to the above decomposition of H_0 into fields, such that

$$K_i = e_i H_0 = H_0 e_i = e_i H_0 e_i \quad \text{for each } i \ (1 \leqq i \leqq m).$$

Then

$$e_i H^2_{sp}(S,\mathbb{Q})$$

is a K_i-module of rank 4. Since the action of H keeps the Hodge type of any element of $H^2_{sp}(S,\mathbb{Q})$, $e_i H^2_{sp}(S,\mathbb{Q})$ is a Hodge substructure of $H^2_{sp}(S,\mathbb{Q})$ for each i. Moreover, since

$$\Psi_S(T^*_{\sigma}(\delta_1), \delta_2) = \Psi_S(\delta_1, T^*_{\sigma}(\delta_2))$$

for any element T_{σ} of H (cf. Section 1.11), we have

$$\Psi_S(e_i H^2_{sp}(S,\mathbb{Q}), e_j H^2_{sp}(S,\mathbb{Q})) = \Psi_S(e_i e_j H^2_{sp}(S,\mathbb{Q}), H^2_{sp}(S,\mathbb{Q})) = 0,$$

if $e_i \neq e_j$. Here Ψ_S is the polarization

$$\Psi_S : H^2_{sp}(S,\mathbb{Q}) \times H^2_{sp}(S,\mathbb{Q}) \longrightarrow \mathbb{Q}(-2).$$

Therefore we have a decomposition as polarized Hodge structures

$$H^2_{sp}(S,\mathbb{Q}) = \bigoplus_{i=1}^{m} e_i H^2_{sp}(S,\mathbb{Q}),$$

where the polarization of each $e_i H^2_{sp}(S,\mathbb{Q})$ is defined by the restriction of Ψ_S. The field $K_i = e_i H_0 e_i$ acts on $e_i H^2_{sp}(S,\mathbb{Q})$. We denote this action by

$$\theta^*_i : K_i \hookrightarrow \mathrm{End}(e_i H^2_{sp}(S,\mathbb{Q})).$$

2.3. Let us consider the Hodge decomposition of each $e_i H^2_{sp}(S,\mathbb{Q}) \otimes_{\mathbb{Q}} \mathbb{C}$. Under the notation of Theorem 1.12, $e_i H^{2,0}$, $e_i H^{1,1}$, and $e_i H^{0,2}$ are the

(2,0) type, (1,1) type, and (0,2) type components of $e_i H^2_{sp}(S,\mathbb{Q})\otimes_{\mathbb{Q}}\mathbb{C}$, respectively. Since the action of K_i holds Hodge type of the elements of $e_i H^2_{sp}(S,\mathbb{Q})\otimes_{\mathbb{Q}}\mathbb{C}$, $K_i\otimes_{\mathbb{Q}}\mathbb{C}$ acts on $e_i H^{2,0}$ by restriction.

We can identify the space $H^{2,0}$ with $S_2(SL_2(O_F)$ as H-modules by Theorem 1.12. Therefore $e_i H^{2,0}$ is a H-submodule of $SL_2(SL_2(O_F))$. Since $H^{2,0}$ is a free $H_0\otimes_{\mathbb{Q}}\mathbb{C}$-module of rank 1, $e_i H^{2,0}$ is also a free $K_i\otimes_{\mathbb{Q}}\mathbb{C}$-module of rank 1. Since

$$K_i\otimes_{\mathbb{Q}}\mathbb{C} = \underbrace{\mathbb{C}\oplus\ldots\oplus\mathbb{C}}_{d_i} \quad \text{with } d_i=[K_i:\mathbb{Q}],$$

we have the corresponding decomposition of $e_i H^{2,0}$

$$e_i H^{2,0} = \mathbb{C}\omega_{f_1}\oplus\mathbb{C}\omega_{f_2}\oplus\ldots\oplus\mathbb{C}\omega_{f_{d_i}}, \quad \text{with } f_j\in S_2(SL_2(O_F))\ (1\leqq j\leqq d_i).$$

Thus for each j $(1\leqq j\leqq d_i)$ and any element $a\in K_i$, we have

$$\theta_i^*(a)\omega_{f_j} = a_j\omega_{f_j} \quad \text{for some } a_j\in\mathbb{C}.$$

Hence, the correspondence

$$\phi_j : a\in K_i \longmapsto a_j\in\mathbb{C}$$

defines an embedding of K_i into $\mathbb{C}$ for each j $(1\leqq j\leqq d_i)$. The action of the Hecke ring H on $e_i H^{2,0}$ factors through $K_i\otimes_{\mathbb{Q}}\mathbb{C}$, thus H acts as scalar multiple on each ω_{f_j}. Therefore f_j are primitive forms, and the image of K_j by ϕ_j is the field K_{f_j} of eigenvalues of f_j. Moreover the set of embeddings ϕ_j $(1\leqq j\leqq d_i)$ gives all the embeddings of K_i into $\mathbb{C}$. In fact, since $d_i=[K_i:\mathbb{Q}]$, K_i has at most d_i different embeddings into $\mathbb{C}$. And $\phi_j=\phi_{j'}$ implies that f_j is a constant multiple of $f_{j'}$ by Theorem 2.2, accordingly $\dim_{\mathbb{C}} e_i H^{2,0}\leqq d_i$. Thus for k $(1\leqq k\leqq m)$ $\dim_{\mathbb{C}} e_k H^{2,0}\leqq d_k$. Therefore,

$$p_g=\dim_{\mathbb{C}} H^{2,0}=\sum_{s=1}^{m}\dim_{\mathbb{C}} e_s H^{2,0} \leqq\sum_{s=1}^{m} d_s = p_g,$$

hence $\dim_{\mathbb{C}} e_s H^{2,0}=d_s$.

2.4. For each primitive idempotent e_i of H_0, we can find d_i primitive forms f_j $(1\leqq j\leqq d_i)$. To make the dependency of f_j on e_i clear, we denote by f_j by f_{ij}. Then the primitive forms $\{f_{ij} \mid 1\leqq i\leqq m, 1\leqq j\leqq d_i\}$ spans the space $S_2(SL_2(O_F))$, and any primitive form of $S_2(SL_2(O_F))$ is a constant multiple of some f_{ij}. Therefore for a given primitive form f of $S_2(SL_2(O_F))$, we can find a primitive idempotent e_i such that

$\omega_f \in e_i H^{2,0}$. We denote this e_i by e_f.

2.5. Definition. For a primitive form $f \in S_2(SL_2(O_F))$, we define a polarized rational Hodge structure $H^2(M_f,\mathbb{Q})$ by

$$H^2(M_f,\mathbb{Q}) = e_f H^2_{sp}(S,\mathbb{Q}).$$

Moreover via the isomorphism

$$\phi_f : e_f H_0 e_f \xrightarrow{\sim} K_f,$$

we can define an action of K_f on $H^2(M_f,\mathbb{Q})$. We denote this action by

$$\theta_f^* : K_f \hookrightarrow \mathrm{End}(H^2(M_f,\mathbb{Q})).$$

Restricting the polarization ψ_S of $H^2_{sp}(S,\mathbb{Q})$ to a direct factor $H^2(M_f,\mathbb{Q})$, we define a polarization Φ_f of $H^2(M_f,\mathbb{Q})$

$$\Phi_f : H^2(M_f,\mathbb{Q}) \times H^2(M_f,\mathbb{Q}) \longrightarrow \mathbb{Q}(-2),$$

which satisfies

$$\Phi_f(\theta_f^*(a)\delta_1,\ \delta_2) = \Phi_f(\delta_1,\ \theta_f^*(a)\delta_2)$$

for any $a \in K_f$, and any $\delta_1,\ \delta_2 \in H^2(M_f,\mathbb{Q})$.

Remark. The symbol M_f is simply an index, and has no more meaning.

2.6. Let f be a primitive form of $S_2(SL_2(O_F))$, and let $a_{\mathfrak{a}}$ be the eigenvalue of f with respect to $T_{\mathfrak{a}}$ for each ideal $\mathfrak{a}$. Suppose that $\sigma : K_f \longrightarrow \mathbb{C}$ be an embedding of K_f into $\mathbb{C}$. Then by the arguement of Paragraph 2.4, we can find a primitive form f^σ unique up to constant multiple, such that

$$T_{\mathfrak{a}}(f^\sigma) = a_{\mathfrak{a}} f^\sigma$$

for any Hecke operator $T_{\mathfrak{a}}$. Moreover we have $e_f = e_{f^\sigma}$ for such f^σ.

Conversely, if $e_f = e_g$ for two primitive forms f, g of $S_2(SL_2(O_F))$, we can find an isomorphism

$$\sigma : K_f \xrightarrow{\sim} K_g,$$

such that

$$T_{\mathfrak{a}}(f) = a_{\mathfrak{a}} f, \text{ and } T_{\mathfrak{a}}(g) = a_{\mathfrak{a}}^\sigma g \quad \text{for any } T_{\mathfrak{a}} \in H.$$

A primitive form g of $S_2(SL_2(O_F))$ is called a companion of a primitive form $f \in S_2(SL_2(O_F))$, if $e_f = e_g$. Let Ξ be a maximal subset of the set of all primitive forms of $S_2(SL_2(O_F))$, such that any two elements of Ξ are not companions mutually. Then

$$H_0 \xrightarrow{\sim} \bigoplus_{f \in \Xi} K_f, \text{ and } H^2_{sp}(S,\mathbb{Q}) = \bigoplus_{f \in \Xi} H^2(M_f,\mathbb{Q}).$$

§3. Nonholomorphic involutive automorphisms of Hilbert modular surfaces.

In this section, we define certain nonholomorphic involutive automorphisms of a Hilbert modular surface, and investigate their actions on the homology groups and cohomology groups. Main statement of this section is Construction 3.7.

3.1. Let ε be a unit of O_F such that $\varepsilon>0$ and $\varepsilon'<0$. Let G_∞, H_∞, and F_∞ be automorphisms of $H\times H$ defined by

$$G_\infty:(z_1, z_2)\in H\times H \longmapsto (\varepsilon z_1, \varepsilon'\bar{z}_2)\in H\times H,$$

$$H_\infty:(z_1, z_2)\in H\times H \longmapsto (\varepsilon'\bar{z}_1, \varepsilon z_2)\in H\times H,$$

$$F_\infty:(z_1, z_2)\in H\times H \longmapsto (-\bar{z}_1, -\bar{z}_2)\in H\times H,$$

respectively. Clearly we have $G_\infty H_\infty=H_\infty G_\infty=F_\infty$. It is easy to check that these automorphisms are normalizers of $SL_2(O_F)$. Therefore, on passing to the quotient S, we have involutive automorphisms of S, which we denote by the same symbols

$$G_\infty:S\longrightarrow S,\quad H_\infty:S\longrightarrow S,\quad \text{and } F_\infty:S\longrightarrow S.$$

Naturally these automorphisms act on the homology groups and cohomology groups $H_2(S,\mathbb{Q})$, $H^2(S,\mathbb{Q})$, $\tilde{H}_2(S,\mathbb{Q})$, $H^2(\partial S_M,\mathbb{Q})$, and $W_2H^2(S,\mathbb{Q})$. Since $G^*_\infty(\eta_i)$, $H^*_\infty(\eta_i)$, and $F^*_\infty(\eta_i)$ $(i=1,2)$ are $+\eta_i$ or $-\eta_i$, these also act on $H_2^{sp}(S,\mathbb{Q})$ and $H^2_{sp}(S,\mathbb{Q})$. Note here that G_∞ and H_∞ change the orientation of the variety S, which we can check by applying G_∞ and H_∞ to $dz_1\wedge d\bar{z}_1\wedge dz_2\wedge d\bar{z}_2$.

3.2. Remark. The automorphism F_∞ is "the Frobenius at the infinite place". The surface S has the canonical model over $\mathbb{Q}$, a fortiori over $\mathbb{R}$. If we write $S_\mathbb{R}$ for the canonical model of S defined over $\mathbb{R}$, then F_∞ coincides with the action of the nontrivial element of $\mathrm{Gal}(\mathbb{C}/\mathbb{R})$ on the $\mathbb{C}$-valued points S of $S_\mathbb{R}$ (cf. Deligne [12]).

3.3. It is easy to check that the action of G_∞, H_∞, and F_∞ on $H_2^{sp}(S,\mathbb{Q})$ and $H^2_{sp}(S,\mathbb{Q})$ commute with that of Hecke operators. Therefore, by restriction, G_∞, H_∞, and F_∞ acts on $H_2(M_f,\mathbb{Q})=e_fH_2^{sp}(S,\mathbb{Q})$ and $H^2(M_f,\mathbb{Q})$.

Now let s and s' be two variable symbols which take the values "+" and "-". Then we put

$$H_2(M_f,\mathbb{Q})_{ss'}=\{\ \gamma\in H_2(M_f,\mathbb{Q})\ |\ G_\infty(\gamma)=s\gamma,\ H_\infty(\gamma)=s'\gamma\ \},$$

and

$$H^2(M_f,\mathbb{Q})_{ss'}=\{\ \delta\in H^2(M_f,\mathbb{Q})\ |\ G_\infty^*\delta=s\delta,\ H_\infty^*\delta=s'\delta\ \}.$$

Clearly we have a decomposition

$$H_2(M_f,\mathbb{Q})=H_2(M_f,\mathbb{Q})_{++}\oplus H_2(M_f,\mathbb{Q})_{+-}\oplus H_2(M_f,\mathbb{Q})_{-+}\oplus H_2(M_f,\mathbb{Q})_{--},$$

and similarly for cohomology. Moreover we have the following proposition on the intersection form Φ_f on $H_2(M_f,\mathbb{Q})$.

3.4. Proposition. *Let f be a primitive form of* $S_2(SL_2(O_F))$. *Then each* $H_2(M_f,\mathbb{Q})_{ss'}$ *is of rank 1 over* K_f *for any* s *and* s'. *Moreover, if we denote by* Φ_f *the intersection form over* $H_2(M_f,\mathbb{Q})$, *then*

$$\Phi_f(\gamma_1,\ \gamma_2)=0 \text{ for any } \gamma_1\in H_2(M_f,\mathbb{Q})_{s_1s_1'}, \text{ and } \gamma_2\in H_2(M_f,\mathbb{Q})_{s_2s_2'},$$

if $s_1=s_2$ *or* $s_1'=s_2'$.

The restriction of Φ_f *to* $H_2(M_f,\mathbb{Q})_{++}\times H_2(M_f,\mathbb{Q})_{--}$ *or to* $H_2(M_f,\mathbb{Q})_{+-}\times H_2(M_f,\mathbb{Q})_{-+}$ *defines a perfect pairing. Especially the bilinear form* Φ_f *is a kernel form.*

Proof. Since G_∞ and H_∞ change the orientation, we have

$$\Phi_f(\gamma,\ \gamma')=-\Phi_f(G_\infty(\gamma),\ G_\infty(\gamma'))=-\Phi_f(H_\infty(\gamma),\ H_\infty(\gamma')),$$

for any γ, γ' of $H_2(M_f,\mathbb{Q})$.

Therefore, if $G_\infty(\gamma)=s\gamma$, and $G_\infty(\gamma')=s\gamma'$ for some γ, γ' of $H_2(M_f,\mathbb{Q})$ with $s=+$, or $s=-$, then $\Phi_f(\gamma,\ \gamma')=-\Phi_f(\gamma,\ \gamma')=0$. We can discuss the case of H_∞ similarly. Hence the second statement of our proposition follows. The third statement is an immediate consequence of this and nondegeneracy of Φ_f. Let us check the first statement. Since the action of Hecke operators and that of G_∞ and H_∞ commute, each $H_2(M_f,\mathbb{Q})_{ss'}$ is a K_f-module. In order to show that $H_2(M_f,\mathbb{Q})_{ss'}$ is of rank 1 over K_f, on passing to the dual space and by extension of scalars, it suffices to check that $H^2(M_f,\mathbb{Q})_{ss'}\otimes_{\mathbb{Q}}\mathbb{C}$ is a $K_f\otimes_{\mathbb{Q}}\mathbb{C}$-module of rank 1. Since $G_\infty^*(\omega_f)=\eta_{f,1}$, $H^*(\omega_f)=\eta_{f,2}$, and $F^*(\omega_f)=c_f\overline{\omega}_f$ with some constant c_f for any primitive form f, we have

$$H^2(M_f,\mathbb{Q})_{++}\otimes_{\mathbb{Q}}\mathbb{C}=\bigoplus_{\sigma:K_f\hookrightarrow\mathbb{C}}\mathbb{C}(\omega_{f\sigma}+\eta_{f\sigma,1}+\eta_{f^\sigma,2}+c_{f\sigma}\overline{\omega}_{f\sigma}),$$

where f^σ are companions of f. Hence $H^2(M_f,\mathbb{Q})_{++}\otimes_{\mathbb{Q}}\mathbb{C}$ is of rank 1 over $K_f\otimes_{\mathbb{Q}}\mathbb{C}$. Discussing other cases similarly, we can show our proposition. q.e.d.

3.5. Corollary. *The intersection form* Φ_f *splits into a composition*

$$H^2(M_f,\mathbb{Q})\times H^2(M_f,\mathbb{Q})\xrightarrow[\psi_f]{}K_f\xrightarrow[\tau]{}\mathbb{Q}.$$

Here ψ_f is a K_f-bilinear form, and τ is a $\mathbb{Q}$-linear mapping given by the trace

$$\tau(x)=tr_{K_f/\mathbb{Q}}(x) \quad \text{for } x\in K_f,$$

with respect to the extension $K_f/\mathbb{Q}$.
Moreover ψ_f is a kernel form over K_f, i.e. of index 2 over K_f.
Similar result is also valid for $H^2(M_f,\mathbb{Q})$.
Proof. Let us denote the action of K_f on $H_2(M_f,\mathbb{Q})$ by

$$\theta_f: K_f \hookrightarrow End(H_2(M_f,\mathbb{Q})).$$

Then any elements γ, γ' of $H_2(M_f,\mathbb{Q})$ are written as linear combinations

$$\gamma = \theta(a)\gamma_{++} + \theta(b)\gamma_{+-} + \theta(c)\gamma_{-+} + \theta(d)\gamma_{--},$$

$$\gamma' = \theta(a')\gamma_{++} + \theta(b')\gamma_{+-} + \theta(c')\gamma_{-+} + \theta(d')\gamma_{--},$$

with a, b, c, d, a', b', c', d' $\in K_f$, with respect to some generators $\gamma_{ss'}$ of $H_2(M_f,\mathbb{Q})_{ss'}$ over K_f.
Hence,

$$\begin{aligned}\Phi_f(\gamma, \gamma') &= \Phi_f(\theta(a)\gamma_{++}, \theta(d')\gamma_{--}) + \Phi_f(\theta(b)\gamma_{+-}, \theta(c')\gamma_{-+}) \\ &\quad + \Phi_f(\theta(c)\gamma_{-+}, \theta(b')\gamma_{-+}) + \Phi_f(\theta(d)\gamma_{--}, \theta(a')\gamma_{++}) \\ &= \Phi_f(\theta(ad')\gamma_{++}, \gamma_{--}) + \Phi_f(\theta(bc')\gamma_{+-}, \gamma_{-+}) \\ &\quad + \Phi_f(\theta(b'c)\gamma_{-+}, \gamma_{+-}) + \Phi_f(\theta(a'd)\gamma_{--}, \gamma_{++}) \\ &= \Phi_f(\theta(ad'+da')\gamma_{++}, \gamma_{--}) + \Phi_f(\theta(bc'+cb')\gamma_{+-}, \gamma_{-+}).\end{aligned}$$

A mapping $K \longrightarrow \mathbb{Q}$ given by

$$a'' \longmapsto \Phi_f(\theta(a'')\gamma_{++}, \gamma_{--})$$

is a linear mapping. Therefore there exists an element $t_1 \in K_f$ such that $\Phi_f(\theta(a'')\gamma_{++}, \gamma_{--})=tr_{K_f/\mathbb{Q}}(t_1 a'')$ for all $a''\in K_f$.

Thus we can find elements t_1 and t_2 of K_f such that

$$\Phi_f(\gamma, \gamma')=tr_{K_f/\mathbb{Q}}(t_1(ad'+da')+t_2(bc'+cb')).$$

Define ψ_f by

$$\psi_f(\gamma, \gamma')=t_1(ad'+da')+t_2(bc'+cb').$$

Then $\Phi_f(\gamma, \gamma')=tr_{K_f/\mathbb{Q}}(\psi_f(\gamma, \gamma'))$, and ψ_f is K_f-bilinear.
By its construction, ψ_f is clearly a kernel form over K_f. q.e.d.
Remark. It is easy to check that ψ_f is uniquely determined, not depending on the choice of generators $\gamma_{ss'}$.

3.6. Let f be a primitive form in $S_2(SL_2(O_F))$, and let K_f be the field of eigenvalues of f. Then K_f is a totally real number field. In fact, by using the Petersson metric of $S_2(SL_2(O_F))$, we can show that K_f is a a subfield of $\mathbb{R}$. Moreover, for any embedding $\sigma:K_f \hookrightarrow \mathbb{C}$, the image $\sigma(K_f)$ is also the field of eigenvalues of the companion f^σ of f. Hence $\sigma(K_f)$ is also a subfield of $\mathbb{R}$. Therefore, K_f is totally real.

Put $d=[K_f:\mathbb{Q}]$, and let $\{\sigma_1,\ldots,\sigma_d\}$ be the set of all embeddings of the totally real number field K_f into $\mathbb{R}$. We denote the composition

$$K_f \xrightarrow{\sigma_i} \mathbb{R} \hookrightarrow \mathbb{C}$$

of each $\sigma_i:K_f \hookrightarrow \mathbb{R}$ with the embedding $\mathbb{R}\hookrightarrow\mathbb{C}$ by the same symbol σ_i. For each σ_i, we denote by f_i a companion of f corresponding to σ_i. The primitive form f_i is unique up to constant multiple with $K_{f_i}=\sigma_i(K_f)$. We denote by $H^2(M_f,\mathbb{Q})\otimes_{K_f,\sigma_i}\mathbb{C}$ the extension of scalars of $H^2(M_f,\mathbb{Q})$ via σ_i as K_k-module. Then we have a Hodge decomposition

$$H^2(M_f,\mathbb{Q})\otimes_{K_f,\sigma_i}\mathbb{C} = \mathbb{C}\omega_{f_i} \oplus \mathbb{C}\eta_{f_i,1} \oplus \mathbb{C}\eta_{f_i,2} \oplus \mathbb{C}\overline{\omega}_{f_i}.$$

Let us consider the Tate twist $H^2(M_f,\mathbb{Q})(1)$ of the Hodge structure $H^2(M_f,\mathbb{Q})$ attached to f. Then we have the following.

3.7. <u>Construction</u>. (i) <u>$H^2(M_f,\mathbb{Q})(1)$ is a Hodge structure of weight 0 of rank 4d with a K_f-action $\theta_f^*:K_f \hookrightarrow \mathrm{End}(H^2(M_f,\mathbb{Q})(1))$. Moreover, as a K_f-module, $H^2(M_f,\mathbb{Q})(1)$ is of rank 4.</u>

(ii) <u>There is a K_f-bilinear form</u>

$$\psi_f:H^2(M_f,\mathbb{Q})(1) \times H^2(M_f,\mathbb{Q})(1) \longrightarrow K_f,$$

<u>with index 2. Especially for any $\sigma_i:K_f \hookrightarrow \mathbb{R}$ $(1\leq i\leq d)$, the extension of scalars $\psi_f\otimes_{K_f,\sigma_i}\mathbb{R}$ of ψ_f via σ_i over $H^2(M_f,\mathbb{Q})(1)\otimes_{K_f,\sigma_i}\mathbb{R}$ defines a bilinear form of signature (2+,2-).</u>

(iii) <u>We have a natural decomposition</u>

$$H^2(M_f,\mathbb{Q})(1)\otimes_{\mathbb{Q}}\mathbb{C} = \bigoplus_{i=1}^{d} H^2(M_f,\mathbb{Q})(1)\otimes_{K_f,\sigma_i}\mathbb{C}.$$

<u>Each subspace $H^2(M_f,\mathbb{Q})(1)\times_{K_f,\sigma_i}\mathbb{C}$ has a Hodge decomposition with Hodge numbers</u>

$$h^{1,-1}=h^{-1,1}=1,\ h^{0,0}=2.$$

<u>Moreover $\psi_f\otimes_{K_f,\sigma_i}\mathbb{C}$ is a polarization of this Hodge structure of weight 0 for each</u> i $(1\leq i\leq d)$.

§4. Period relation of Riemann-Hodge.

In this section, we present the Riemann-Hodge period relation for the periods of Hilbert modular surfaces, or more precisely speaking, its reformulation for the periods of the Hodge structures $H^2(M_f,\mathbb{Q})$.

4.1. Let f be a normalized primitive form of weight 2, and let K_f be the field of eigenvalues of f. Suppose that $\{\sigma_1,\dots,\sigma_d\}$ is the set of all embeddings of K_f into $\mathbb{R}$, where $d=[K_f:\mathbb{Q}]$. We write f_i in this section for the normalized primitive form which is a companion corresponding to σ_i $(1\leqq i\leqq d)$. We have put

$$\omega_f = (2\pi i)^2 f(z_1, z_2)dz_1\wedge dz_2$$

for any $f\in S_2(SL_2(O_F))$. Then we have the following lemma.

4.2. <u>Lemma</u>. <u>Let</u> f <u>be</u> <u>a</u> <u>normalized</u> <u>primitive</u> <u>form</u> <u>of</u> $S_2(SL_2(O_F))$.

(i) <u>If</u> $\gamma\in H_2(M_f,\mathbb{Q})_{++}$ <u>or</u> $\gamma\in H_2(M_f,\mathbb{Q})_{--}$, <u>then</u> <u>the</u> <u>period</u> <u>integral</u>

$$\int_\gamma \omega_f$$

<u>is</u> <u>a</u> <u>real</u> <u>number</u>.

(ii) <u>If</u> $\gamma\in H_2(M_f,\mathbb{Q})_{+-}$ <u>or</u> $\gamma\in H_2(M_f,\mathbb{Q})_{-+}$, <u>then</u> <u>the</u> <u>period</u> <u>integral</u>

$$\int_\gamma \omega_f$$

<u>is</u> <u>a</u> <u>purely</u> <u>imaginary</u> <u>number</u>.

<u>Proof</u>. If $\gamma\in H_2(M_f,\mathbb{Q})_{++}\oplus H_2(M_f,\mathbb{Q})_{--}$, $F_\infty(\gamma)=\gamma$. Therefore,

$$\int_\gamma \omega_f = \int_{F_\infty(\gamma)} \omega_f = \int_\gamma F_\infty^*(\omega_f).$$

Since every Fourier coefficient of f is a real number, $F^*(\omega_f)=\overline{\omega}_f$
Hence,

$$\int_\gamma \omega_f = \overline{\int_\gamma \omega_f},$$

which implies that

$$\int_\gamma \omega_f$$

is a real number. If $\gamma\in H_2(M_f,\mathbb{Q})_{+-}\oplus H_2(M_f,\mathbb{Q})_{-+}$, then $F_\infty(\gamma)=\gamma$.
Similarly we have

$$\int_\gamma \omega_f = -\overline{\int_\gamma \omega_f}$$

in this case. q.e.d.

4.3. Definition. Let

$$\psi_f: H_2(M_f,\mathbb{Q}) \times H_2(M_f,\mathbb{Q}) \longrightarrow K_f$$

be the K_f-bilinear form defined by Corollary 3.5 of Proposition 3.4. Then, a basis $\{\gamma_{ss'} \mid s=\pm, s'=\pm\}$ over K_f of $H_2(M_f,\mathbb{Q})$ is called a canonical basis, if it satisfies the conditions:

(i) $\gamma_{ss'}$ is a generator of $H_2(M_f,\mathbb{Q})_{ss'}$ over K_f for each (s,s'), and

(ii) $\psi_f(\gamma_{s_1s_1'}, \gamma_{s_2s_2'}) = \begin{cases} +1, & \text{if } s_1\neq s_2, \text{ and } s_1'\neq s_2'. \\ 0, & \text{otherwise.} \end{cases}$

A canonical basis $\{\delta_{ss'} \mid s=\pm, s'=\pm\}$ of $H^2(M_f,\mathbb{Q})$ is defined similarly.

4.4. Theorem. (Period relation of Riemann-Hodge).

Let $\{\gamma_{ss'} \mid s=\pm, s'=\pm\}$ be a canonical basis of $H_2(M_f,\mathbb{Q})$. Put

$$W_{ss'}(f_i) = \int_{\gamma_{ss'}} \omega_{f_i} \quad \text{for each } i \ (1\leq i\leq d).$$

Then, we have

(i) $W_{++}(f_i)W_{--}(f_i) + W_{+-}(f_i)W_{-+}(f_i) = 0$,

and

(ii) $W_{++}(f_i)W_{--}(f_i) = -W_{+-}(f_i)W_{-+}(f_i) = \frac{1}{4}(f_i,f_i)$.

Here (f_i,f_i) is the normalized Petersson metric defined by

$$(f_i,f_i) = \int_S \omega_{f_i} \wedge \overline{\omega}_{f_i}$$

Proof. By Poincaré duality, we have an isomorphism

$$\rho: H_2(M_f,\mathbb{Q}) \cong H^2(M_f,\mathbb{Q}).$$

Because of the uniqueness of ψ_f, the set of the images $\delta_{ss'}=\rho(\gamma_{ss'})$ of the canonical basis $\{\gamma_{ss'}\}$ also forms a canonical basis of $H^2(M_f,\mathbb{Q})$. Hence,

$$W_{ss'}(f_i) = \int_{\gamma_{ss'}} \omega_{f_i} = (\Phi_f \otimes_{\mathbb{Q}} \mathbb{C})(\delta_{ss'}, \omega_{f_i}).$$

Let $\delta^i_{ss'}$ be the image of $\delta_{ss'}$ in $H^2(M_f,\mathbb{Q})\otimes_{K_f,\sigma_i}\mathbb{C}$ for each i and any (s,s'). Then in view of the decompositions

$$H^2(M_f,\mathbb{Q})\otimes_{\mathbb{Q}}\mathbb{C} = \bigoplus_{j=1}^{d} H^2(M_f,\mathbb{Q})\otimes_{K_f,\sigma_j}\mathbb{C}$$

and

$$\Phi_f\otimes_{\mathbb{Q}}\mathbb{C} = \mathrm{tr}_{K_f/\mathbb{Q}}(\psi_f)\otimes_{\mathbb{Q}}\mathbb{C} = \bigoplus_{j=1}^{d} \psi_f\otimes_{K_f,\sigma_j}\mathbb{C},$$

we have

$$W_{ss'}(f_i)=\sum_{j=1}^{d} (\psi_f\otimes_{K_f,\sigma_j}\mathbb{C})(\delta^j_{ss'},\ \omega_{f_i}).$$

Since

$$(\psi_f\otimes_{K_f,\sigma_j}\mathbb{C})(\delta^j_{ss'},\ \omega_{f_i})=0,\ \text{if } i\neq j,$$

$$W_{ss'}(f_i)= (\psi_f\otimes_{K_f,\sigma_i}\mathbb{C})(\delta^i_{ss'},\ \omega_{f_i}).$$

Therefore,

$$\omega_{f_i}=W_{--}(f_i)\delta^i_{++} + W_{-+}(f_i)\delta^i_{+-} + W_{+-}(f_i)\delta^i_{-+} + W_{++}(f_i)\delta^i_{--}.$$

for each i ($1\leq i\leq d$).

Consider the intersection numbers

$$(\Phi_f\otimes_{\mathbb{Q}}\mathbb{C})(\omega_i,\ \omega_i) \text{ and } (\Phi_f\otimes_{\mathbb{Q}}\mathbb{C})(\omega_i,\ \overline{\omega}_i).$$

Applying Lemma 4.2 and de Rham's theorem, we have

$$(\Phi_f\otimes_{\mathbb{Q}}\mathbb{C})(\omega_{f_i},\ \omega_{f_i})=\int_S \omega_i\wedge\omega_i$$

$$= 2W_{++}(f_i)W_{--}(f_i) + 2W_{+-}(f_i)W_{-+}(f_i),$$

and

$$(\Phi_f\otimes_{\mathbb{Q}}\mathbb{C})(\omega_{f_i},\ \overline{\omega}_{f_i})=\int_S \omega_i\wedge\overline{\omega}_i$$

$$= 2W_{++}(f_i)W_{--}(f_i) - 2W_{+-}(f_i)W_{-+}(f_i).$$

Since $\omega_i\wedge\omega_i=0$ for any i, these equalities imply immediately our theorem. q.e.d.

4.5. Remark. The system of periods $\{ W_{ss'}(f_i) \mid s=\pm,\ s'=\pm,\ 1\leq i\leq d \}$ depends on the choice of the canonical basis $\{\gamma_{ss'}\}$. Another possible choice of a canonical basis is given by

$$\{ \theta(a)\gamma_{++},\ \theta(b)\gamma_{+-},\ \theta(b^{-1})\gamma_{-+},\ \theta(a^{-1})\gamma_{--} \}$$

for some a, $b\in K_f^{\times}$.

Then the system of periods changes to

$$\{ \sigma_i(a)W_{++}(f_i),\ \sigma_i(b)W_{+-}(f_i),\ \sigma_i(b^{-1})W_{-+}(f_i),\ \sigma_i(a^{-1})W_{--}(f_i) \mid 1\leq i\leq d \}.$$

Chapter II. Abelian varieties attached to primitive forms.

§5. Abelian varieties attached to Clifford algebras.

Starting with a datum of Hodge structure given in Section 3.7, in this section we construct an abelian variety. The method of this construction is a slight generalization of that of Satake [40], Kuga-Satake [20], and Deligne [10]. More precisely speaking, we consider the case where the given Hodge structure has a totally real algebraic number field as an endomorphism ring, in place of the trivial one $\mathbb{Q}$.

Applying this construction to the Hodge structures $H^2(M_f,\mathbb{Q})$ attached to primitive Hilbert modular cusp forms f of weight 2, we can attach an abelian variety A(f) of dimension 4d to any primitive form f of $S_2(SL_2(O_F))$ with $[K_f:\mathbb{Q}]=d$. The existence of non-holomorphic involutive automorphisms G_∞, H_∞, and F_∞ implies that A(f) is isogenous to a product $A_f^1\times A_f^1\times A_f^2\times A_f^2$ of two kinds of abelian varieties A_f^1 and A_f^2 of dimension d with endomorphism algebra K_f.

5.1. We start with the following datum.

Datum. (i) A totally real algebraic number field K of degree d over $\mathbb{Q}$, and a rational Hodge structure $H_\mathbb{Q}$ of weight 0 with a K-biliear form

$$\psi: H_\mathbb{Q}\times H_\mathbb{Q} \longrightarrow K.$$

Moreover the elements of K act on $H_\mathbb{Q}$ as endomorphisms of Hodge structures. The rank of $H_\mathbb{Q}$ over K is q+2.

(ii) Let $\{\sigma_1,\ldots,\sigma_d\}$ be the set of all embeddings of K into $\mathbb{R}$ (or into $\mathbb{C}$ by the same symbols). And suppose that

$$H_\mathbb{Q}\otimes_\mathbb{Q}\mathbb{R} = \bigoplus_{i=1}^{d} H_\mathbb{Q}\otimes_{K,\sigma_i}\mathbb{R},$$

and

$$H_\mathbb{Q}\otimes_\mathbb{Q}\mathbb{C} = \bigoplus_{i=1}^{d} H_\mathbb{Q}\otimes_{K,\sigma_i}\mathbb{C},$$

be the canonical decompositions into eigenspaces with respect to the action of K. Then, for each σ_i $(1\leqq i\leqq d)$, the extension of scalars

$\psi\otimes_{\sigma_i}\mathbb{R}$ of ψ is of signature (q+,2-), and defines a polarization of the Hodge structure $H_{\mathbb{Q}}\otimes_{K,\sigma_i}\mathbb{R}$.

(iii) The Hodge numbers of each $H_{\mathbb{Q}}\otimes_{K,\sigma_i}\mathbb{R}$ are given by

$$h^{1,-1}=h^{-1,1}=1 \quad \text{and} \quad h^{0,0}=q.$$

5.2. Let $SO(H_{\mathbb{Q}})$ be the algebraic orthogonal group defined over K with respect to the bilinear form ψ. Consider the restriction of scalars $Res_{K/\mathbb{Q}}(SO(H_{\mathbb{Q}}))$ of $SO(H_{\mathbb{Q}})$ with respect to the extension $K/\mathbb{Q}$. Then the set of real points of this algebraic group over $\mathbb{Q}$ is canonically isomorphic to

$$SO(H_{\mathbb{Q}}\otimes_{\mathbb{Q}}\mathbb{R})=\prod_{i=1}^{d} SO(H_{\mathbb{Q}}\otimes_{K,\sigma_i}\mathbb{R}).$$

In terms of the formulation of Deligne [10], our datum defines a homomorphism

$$h:\underline{S}=Res_{\mathbb{C}/\mathbb{R}}(\mathbb{G}_m) \longrightarrow SO(H_{\mathbb{Q}}\otimes_{\mathbb{Q}}\mathbb{R})=\prod_{i=1}^{d} SO(H_{\mathbb{Q}}\otimes_{K,\sigma_i}\mathbb{R})$$

of weight 0.

Let us recall the formalism of [10], slightly modifying it. We denote by $C(H_{\mathbb{Q}})$ or by $C(H_{\mathbb{Q}},\psi)$ (resp. $C^+(H_{\mathbb{Q}})$ or by $C^+(H_{\mathbb{Q}},\psi)$) the Clifford algebra (resp. the even Clifford algebra) of K-module $H_{\mathbb{Q}}$ over K with respect to ψ. We denote by $CSpin(H_{\mathbb{Q}})$ the Clifford group.

A natural homomorphism

$$CSpin(H_{\mathbb{Q}}) \longrightarrow SO(H_{\mathbb{Q}})$$

of algebraic groups over K induces a homomorphism

$$Res_{K/\mathbb{Q}}CSpin(H_{\mathbb{Q}}) \longrightarrow Res_{K/\mathbb{Q}}SO(H_{\mathbb{Q}})$$

of algebraic groups over $\mathbb{Q}$.

Let $C^+(H_{\mathbb{Q}})^*$ be the set of invertible elements of the algebra $C^+(H_{\mathbb{Q}})$. Then, by definition, we have a natural inclusion mapping

$$\alpha:CSpin(H_{\mathbb{Q}}) \hookrightarrow C^+(H_{\mathbb{Q}})^*.$$

By using this α, we define a representation $C^+(H_{\mathbb{Q}})_s$ of the K-algebraic group $CSpin(H_{\mathbb{Q}})$ on $C^+(H_{\mathbb{Q}})$ by means of

$$g*x=\alpha(g)\cdot x, \quad \text{for } x\in C^+(H_{\mathbb{Q}}).$$

Applying the functor $Res_{K/\mathbb{Q}}$ of restriction of scalars, we have a representation $Res_{K/\mathbb{Q}}(C^+(H_{\mathbb{Q}})_s)$ over $\mathbb{Q}$ of $\mathbb{Q}$-algebraic group

$\mathrm{Res}_{K/\mathbb{Q}}\mathrm{CSpin}(H_{\mathbb{Q}})$ on the $\mathbb{Q}$-vector space $\mathrm{Res}_{K/\mathbb{Q}}(C^{+}(H_{\mathbb{Q}}))$.

Similarly we can define a representation $\mathrm{Res}_{K/\mathbb{Q}}(C^{+}(H_{\mathbb{Q}})_{ad})$ over $\mathbb{Q}$ of $\mathrm{Res}_{K/\mathbb{Q}}\mathrm{CSpin}(H_{\mathbb{Q}})$ on $\mathrm{Res}_{K/\mathbb{Q}}(C^{+}(H_{\mathbb{Q}}))$, via the representation $C^{+}(H_{\mathbb{Q}})_{ad}$ over K of $\mathrm{CSpin}(H_{\mathbb{Q}})$ over $C^{+}(H_{\mathbb{Q}})$ defined by

$$g_{ad}*x = \alpha(g)x\alpha(g)^{-1}.$$

We can regard $C^{+}(H_{\mathbb{Q}})_s$ as $C^{+}(H_{\mathbb{Q}})$-module by right multiplication. Then the action of $\mathrm{CSpin}(H_{\mathbb{Q}})$ on $C^{+}(H_{\mathbb{Q}})_s$ commutes with the right action of $C^{+}(H_{\mathbb{Q}})$, and moreover we have an isomorphism of representation of $\mathrm{CSpin}(H_{\mathbb{Q}})$ over K

$$\underline{\mathrm{End}}_{C^{+}(H_{\mathbb{Q}})}(C^{+}(H_{\mathbb{Q}})_s) \cong C^{+}(H_{\mathbb{Q}})_{ad},$$

which implies that there is an isomorphism of the representations of $\mathrm{Res}_{K/\mathbb{Q}}\mathrm{CSpin}(H_{\mathbb{Q}})$ over $\mathbb{Q}$

$$\underline{\mathrm{End}}_{\mathrm{Res}_{K/\mathbb{Q}}C^{+}(H_{\mathbb{Q}})}(\mathrm{Res}_{K/\mathbb{Q}}C^{+}(H_{\mathbb{Q}})_s) \cong \mathrm{Res}_{K/\mathbb{Q}}C^{+}(H_{\mathbb{Q}})_{ad}.$$

5.3. By [10], the homomorphism

$$h:\underline{S} \longrightarrow \mathrm{Res}_{K/\mathbb{Q}}SO(H_{\mathbb{Q}})(\mathbb{R}) = SO(H_{\mathbb{Q}}\otimes_{\mathbb{Q}}\mathbb{R}) = \prod_{i=1}^{d} SO(H_{\mathbb{Q}}\otimes_{K,\sigma_i}\mathbb{R})$$

defined by our datum 5.1, is lifted in a unique way to a homomorphism

$$\tilde{h}:\underline{S} \longrightarrow \mathrm{Res}_{K/\mathbb{Q}}\mathrm{CSpin}(H_{\mathbb{Q}})(\mathbb{R}) = \prod_{i=1}^{d} \mathrm{CSpin}(H_{\mathbb{Q}}\otimes_{K,\sigma_i}\mathbb{R})$$

such that the diagram

$$\begin{array}{ccccc} \mathbb{G}_{m\mathbb{R}} & \xrightarrow{w} & \underline{S} & \xrightarrow{t} & \mathbb{G}_{m\mathbb{R}} \\ \| & & \downarrow \tilde{h}_i & & \| \\ \mathbb{G}_{m\mathbb{R}} & \xrightarrow{w_i} & \mathrm{CSpin}(H_{\mathbb{Q}}\otimes_{K,\sigma_i}\mathbb{R}) & \xrightarrow{t} & \mathbb{G}_{m\mathbb{R}} \end{array}$$

is commutative for each i ($1 \leqq i \leqq d$). Here $\tilde{h} = (\tilde{h}_i)_{1\leqq i\leqq d}$, and

$$w:\mathbb{G}_{m\mathbb{R}} = \mathbb{R}^* \longrightarrow \underline{S}(\mathbb{C}) = \mathbb{C}^*$$

is given by $x \in \mathbb{R}^* \longmapsto x \in \mathbb{C}^*$, and

$$t:\underline{S}(\mathbb{R}) = \mathbb{C}^* \longrightarrow \mathbb{G}_{m\mathbb{R}} = \mathbb{R}^*$$

by $t(z)=(z\bar{z})^{-1}$ for $z \in \mathbb{C}^*$. Each w_i is given by the natural exact sequence

$$0 \longrightarrow \mathbb{G}_{m\mathbb{R}} \xrightarrow{w_i} \mathrm{CSpin}(H_{\mathbb{Q}}\otimes_{K,\sigma_i}\mathbb{R}) \longrightarrow SO(H_{\mathbb{Q}}\otimes_{K,\sigma_i}\mathbb{R}) \longrightarrow 0,$$

and

$$t:\mathrm{CSpin}(H_{\mathbb{Q}}\otimes_{K,\sigma_i}\mathbb{R}) \longrightarrow \mathbb{G}_{m\mathbb{R}}$$

is given by the inverse of the spinor norm mapping

$$\mathbb{N}: CSpin(H_{\mathbb{Q}} \otimes_{K,\sigma_i} \mathbb{R}) \longrightarrow \mathbb{G}_{m\mathbb{R}}.$$

Similarly as [10], every representation of $Res_{K/\mathbb{Q}}CSpin(H_{\mathbb{Q}})$ is polarizable with respect to $\tilde{h}$ (cf. Lemma (4.3) of [10]), and we have the following proposition.

5.4. Proposition. (cf. Proposition (4.5) of [10].)
The Hodge structure of $Res_{K/\mathbb{Q}}(C^+(H_{\mathbb{Q}})_s)$ defined by $\tilde{h}$ is of type $\{(0,1),(1,0)\}$, and polarizable.

Consequently, if we choose a finite type O_K-module $H_{\mathbb{Z}}$, which is a lattice in $H_{\mathbb{Q}}$ such that $H_{\mathbb{Z}} \otimes_{O_K} K = H_{\mathbb{Q}}$, and the restriction $\psi_{\mathbb{Z}}$ of ψ to $H_{\mathbb{Z}} \times H_{\mathbb{Z}}$ takes values in O_K, then the above Hodge structure

$$Res_{K/\mathbb{Q}}(C^+(H_{\mathbb{Q}})_s) \text{ with the lattice } Res_{O_K/\mathbb{Z}}C^+(H_{\mathbb{Z}},\psi_{\mathbb{Z}})$$

defines an abelian variety A. Here O_K is the ring of integers of K. Clearly another choice of lattice $H'_{\mathbb{Z}}$ in $H_{\mathbb{Q}}$ defines an abelian variety isogenous to A. Since

$$rank_{\mathbb{Q}} Res_{K/\mathbb{Q}} C^+(H_{\mathbb{Q}}) = d \cdot rank_K C^+(H_{\mathbb{Q}}) = d \cdot 2^{(q+1)},$$

the dimension of A is $2^q d$.

Put $C^+ = C^+(H_{\mathbb{Z}},\psi_{\mathbb{Z}})$. Then the Hodge structure of the commutant $\underline{End}_{C^+}(A)$ is given by the following proposition due to Deligne [10]. We do not use this proposition later except in remarks.

5.5. Proposition. *There is an isomorphism of Hodge structures compatible with the actions of K*

$$C^+(H_{\mathbb{Q}}) \xrightarrow{\sim} \underline{End}_{C^+}(A) \otimes_{\mathbb{Z}} \mathbb{Q}.$$

Here the Hodge structure of $C^+(H_{\mathbb{Q}})$ is given by $\tilde{h}$ via the representation $Res_{K/\mathbb{Q}}C^+(H_{\mathbb{Q}})_{ad}$.

Remark. Moreover, by Proposition 3 of §9 of [10], we have an isomorphism of Hodge structures with K-actions

$$\bigoplus_i \bigwedge_K^{2i} H_{\mathbb{Q}} \xrightarrow{\sim} C^+(H_{\mathbb{Q}}).$$

5.6. Applying the construction of the previous paragraphs for $K = K_f$, $H_{\mathbb{Q}} = H^2(M_f,\mathbb{Q})(1)$, and $\psi = \psi_f(1)$, we can attach an abelian variety A(f) by choosing a lattice $H_{\mathbb{Z}}$ in $H_{\mathbb{Q}}$. Since q=2 in this case, A(f) is of dimension 4d with $d = [K_f:\mathbb{Q}]$.

The even Clifford algebra $C^+(H_{\mathbb{Z}})$ over $H_{\mathbb{Z}}$ acts on A(f) as endomorphisms by right multiplication. By Corollary 3.5 of Proposition 3.4, ψ_f is a kernel form. Therefore,

$$C^+(H_{\mathbb{Z}})\otimes_{\mathbb{Z}}\mathbb{Q} \cong C^+(H_{\mathbb{Q}}) \cong M_2(K_f) \oplus M_2(K_f)$$

(cf. Bourbaki [7], §9).

Hence A(f) is isogenous to a product

$$A_f^1 \times A_f^1 \times A_f^2 \times A_f^2$$

of two kinds of abelian varieties A_f^1 and A_f^2 of dimension d with an endomorphism algebra

$$K_f \hookrightarrow \mathrm{End}(A_f^i)\otimes_{\mathbb{Z}}\mathbb{Q} \quad (i=1,2).$$

5.7. It is convenient to recall a more naive definition of the complex structure $\tilde{h}(i)_S$ on $(\mathrm{Res}_{K/\mathbb{Q}}C^+(H_{\mathbb{Q}}))\otimes_{\mathbb{Q}}\mathbb{R}$, for an explicit calculation of the period moduli of A_f^1 and A_f^2 in the next section.

Let us recall the definition of the complex structure in Kuga-Satake [20]. First note a natural isomorphism

$$(\mathrm{Res}_{K/\mathbb{Q}}C^+(H_{\mathbb{Q}}))\otimes_{\mathbb{Q}}\mathbb{R} \cong \bigoplus_{i=1}^{d} C^+(H_{\mathbb{Q}})\otimes_{K,\sigma_i}\mathbb{R} \cong \bigoplus_{i=1}^{d} C^+(H_{\mathbb{Q}}\otimes_{K,\sigma_i}\mathbb{R}).$$

To give

$$h=(h_i)_{1\leqq i\leqq d} : \underline{S} \longrightarrow \prod_{i=1}^{d} SO(H_{\mathbb{Q}}\otimes_{K,\sigma_i}\mathbb{R})$$

is equivalent to to give an orientated subspace $H_{\mathbb{R},i}^-$ of dimension 2 in $H_{\mathbb{Q}}\otimes_{K,\sigma_i}\mathbb{R}$, on which $\psi\otimes_{K,\sigma_i}\mathbb{R}$ is negative definite for each i ($1\leqq i\leqq d$). Therefore for each i we can find an orthogonal basis $\{e_i^+, e_i^-\}$ of $H_{\mathbb{R},i}^-$ with respect to $\psi_i=\psi\otimes_{K,\sigma_i}\mathbb{R}$ such that

$$\psi_i(e_i^+, e_i^+)=\psi_i(e_i^-, e_i^-)=-1,\ \psi_i(e_i^+, e_i^-)=0,$$

and $e_i^+\wedge e_i^-$ corresponds to the given orientation.

Put

$$J_i = e_i^+ e_i^- \quad (1\leqq i\leqq d) \text{ for each } i,$$

which is an element of $C^+(H_{\mathbb{Q}}\otimes_{K,\sigma_i}\mathbb{R})$. Then $J=\bigoplus_{i=1}^{d} J_i$ defines a complex structure on $\mathrm{Res}_{K/\mathbb{Q}}C^+(H_{\mathbb{Q}})\otimes_{\mathbb{Q}}\mathbb{R}$, which coincides with $\tilde{h}(i)_S$.

§6. The period moduli of the isogeny classes $A_f^1 \otimes \mathbb{Q}$ and $A_f^2 \otimes \mathbb{Q}$.

In this section, we calculate the period moduli of the abelian varieties A_f^1 and A_f^2, which are constructed in the previous section for a primitive form f of $S_2(SL_2(O_F))$. Since A_f^1 and A_f^2 depend on the choice of a lattice in $H^2(M_f,\mathbb{Q})$, we consider rather their isogeny classes $A_f^1 \otimes \mathbb{Q}$ and $A_f^2 \otimes \mathbb{Q}$.

To formulate our statement, we require some language on the period moduli of Hilbert-Blumenthal abelian varieties.

6.1. Definition. Let K be a totally real algebraic number field of degree d over $\mathbb{Q}$. A Hilbert-Blumenthal abelian variety is a pair of an abelian variety A of dimension d defined over $\mathbb{C}$, and a homomorphism $\theta: K \hookrightarrow \mathrm{End}(A)\otimes_{\mathbb{Z}}\mathbb{Q}$. Let A and B be two Hilbert-Blumenthal abelian varieties with respect to K. Then A and B are said K-isogenous, if there exist some order $\mathcal{O}$ of K and an isogeny $f: A \longrightarrow B$ compatible with the action of $\mathcal{O}$.

6.2. Suppose that A is a Hilbert-Blumenthal abelian variety of dimension d with respect to K. Denote the action of K on the first homology group $H_1(A,\mathbb{Q})$ by θ. Then $H_1(A,\mathbb{Q})$ is a K-module of rank 2 with respect to this action. It is know from the structure theory of abelian varieties (cf. Mumford [27]), that the Rosati involution acts trivially on K. Therefore for the polarization form

$$\Phi_*: H_1(A,\mathbb{Q}) \times H_1(A,\mathbb{Q}) \longrightarrow \mathbb{Q},$$

we have

$$\Phi_*(\theta(a)\gamma_1, \gamma_2)=\Phi_*(\gamma_1, \theta(a)\gamma_2).$$

Therefore, this form Φ_* is written as

$$\Phi_*=\mathrm{tr}_{K/\mathbb{Q}}(\psi),$$

for some skew-symmetric K-bilinear form ψ on $H_1(A,\mathbb{Q})$.

Let $\{\sigma_1,\ldots,\sigma_d\}$ be the set of all embeddings of K into $\mathbb{R}$ or into $\mathbb{C}$. Let

$$\theta^*: K \hookrightarrow \mathrm{End}(H^1(A,\mathbb{Q}))$$

be the natural action of K on the rational cohomology group $H^1(A,\mathbb{Q})$. Then we have a decomposition with respect to this action

$$H^1(A,\mathbb{R}) = H^1(A,\mathbb{Q})\otimes_{\mathbb{Q}}\mathbb{R} = \bigoplus_{i=1}^{d} H^1(A,\mathbb{Q})\otimes_{K,\sigma_i}\mathbb{R}.$$

Decomposing the space $\Gamma(A, \Omega^1_{A/\mathbb{C}})$ of holomorphic 1-forms on A into eigenspaces with respect to K, we have

$$\Gamma(A, \Omega^1_{A/\mathbb{C}}) = \bigoplus_{i=1}^{d} \mathbb{C}\omega_i,$$

where ω_i is characterized by $\theta^*(a)\omega_i = \sigma_i(a)\omega_i$ for any $a \in K$, and unique up to constant multiple.
Thus for each σ_i we have a Hodge decomposition

$$H^1(A,\mathbb{Q}) \otimes_{K,\sigma_i} \mathbb{C} = \mathbb{C}\omega_i \oplus \mathbb{C}\overline{\omega}_i .$$

6.3. Thus, starting with a Hilbert-Blumenthal abelian variety

$$\{ A,\ \theta: K \hookrightarrow \mathrm{End}(A) \otimes_{\mathbb{Z}} \mathbb{Q},\ \Phi \},$$

we have a Hodge structure of weight 1

$$H^1(A,\mathbb{Q}),$$

an action of K on the Hodge structure $H^1(A,\mathbb{Q})$

$$\theta^*: K \hookrightarrow \mathrm{End}(H^1(A,\mathbb{Q})),$$

and a unique skew-symmetric K-bilinear form

$$\psi: H^1(A,\mathbb{Q}) \times H^1(A,\mathbb{Q}) \longrightarrow K$$

such that $\Phi = \mathrm{tr}_{K/\mathbb{Q}}(\psi)$ and the extension of scalars $\psi \otimes_{K,\sigma_i} \mathbb{R}$ is a polarization of the real Hodge structure $H^1(A,\mathbb{Q}) \otimes_{K,\sigma_i} \mathbb{R}$ of weight 1 for each σ_i.

6.4. Choose a K-basis of $\{ \gamma_1, \gamma_2 \}$ of $H^1(A,\mathbb{Q})$ such that

$$\psi(\gamma_1, \gamma_2) = +1,$$

and form period integrals

$$w_i(\omega_j) = \int_{\gamma_i} \omega_j \qquad (i=1,2;\ j=1,\dots,d).$$

Put

$$\tau_j = w_2(\omega_j)/w_1(\omega_j) \quad \text{for each } j \quad (1 \leqq j \leqq d),$$

and define a point τ of $\mathbb{C}^d$ by

$$\tau = (\tau_1, \tau_2, \dots, \tau_d).$$

Then, by the period relation of Riemann, this point belongs to H^d, where H is the complex upper half plane.

Clearly this point does not depends on the choice of ω_i. But it depends on the choice of $\{\gamma_1, \gamma_2\}$. For any other K-basis $\{\gamma_1', \gamma_2'\}$ of $H_1(A,\mathbb{Q})$ with $\psi(\gamma_1', \gamma_2')=1$, there exist four elements a, b, c, d of K with ad-bc=1 such that

$$\gamma_1'=\theta(a)\gamma_1+\theta(b)\gamma_2, \quad \gamma_2'=\theta(c)\gamma_1+\theta(d)\gamma_2.$$

Let us define $w_i'(\omega_j)$ and τ_j' for $\{\gamma_1', \gamma_2'\}$ similarly as for $\{\gamma_1, \gamma_2\}$. Then

$$\tau_j' = \{\sigma_j(d)\tau_j+ \sigma_j(c)\}/\{\sigma_j(b)\tau_j+\sigma_j(a)\} \text{ for each } j.$$

Thus the datum

$$\{ H^1(A,\mathbb{Q}),\ \theta^*:K \hookrightarrow \mathrm{End}(H^1(A,\mathbb{Q})),\ \psi \}$$

defines a point of the orbit space $SL_2(K)\backslash H^d$.
Let us call this orbit τ mod $SL_2(K)$ <u>the period modulus</u> of the datum

$$\{ H^1(A,\mathbb{Q}),\ \theta^*:K \hookrightarrow \mathrm{End}(H^1(A,\mathbb{Q})),\ \psi \}.$$

Thus we can define a mapping from the set of polarized Hilbert-Blumenthal abelian varieties to the set $SL_2(K)\backslash H^d$. In view of the choice of another polarization, we can see that the K-isogeny class of a Hilbert-Blumenthal abelian variety A defines a point of $GL_2(K)\backslash X$, where

$$X=(\mathbb{C}-\mathbb{R})^d,$$

and $GL_2(K)$ acts on X in the usual manner.

The following result is known (cf. Shimura [42] for example).

6.5. <u>Theorem. The above correspondence of the set of the K-isogeny classes of Hilbert-Blumenthal abelian varieties $A\otimes\mathbb{Q}$ with the set of $GL_2(K)$-orbits in X are bijective.</u>

We call this $GL_2(K)$-orbit in X corresponding to $A\otimes\mathbb{Q}$, <u>the period modulus</u> of the K-isogeny class $A\otimes\mathbb{Q}$, and denote it by $\tau(A\otimes\mathbb{Q})$.

Now let us state the main result of this section.

6.6. <u>Theorem. Let f be a primitive form of $S_2(SL_2(O_F))$, and let K_f be the field of eigenvalues of f with the set $\{\sigma_1,\ldots,\sigma_d\}$ of all embeddings of K_f into $\mathbb{C}$. Choose a canonical basis of $H^2(M_f,\mathbb{Q})$, and define the period integrals</u>

$$W_{\mathfrak{s}\mathfrak{s}'}(f_i)$$

<u>as in Theorem 4.4 for the companions f_i of f. Then, the period moduli of the isogeny classes $A_f^1\otimes\mathbb{Q}$ and $A_f^2\otimes\mathbb{Q}$ of the two Hilbert-Blumenthal abelian varieties A_f^1 and A_f^2 constructed in Section 5.6, are given by $GL_2(K_f)$-orbits of the points in X:</u>

$$(W_{+-}(f_1)/W_{++}(f_1),\ldots, W_{+-}(f_d)/W_{++}(f_d))$$

and

$$(W_{-+}(f_1)/W_{++}(f_1),\ldots, W_{-+}(f_d)/W_{++}(f_d)).$$

as a pair.

Proof. First, as in Theorem 4.4, we choose a canonical basis $\{\delta_{ss'}\}$ of $H^2(M_f,\mathbb{Q})$. Let us put

$$\delta_1=\delta_{++},\ \delta_2=\delta_{+-},\ \delta_3=\delta_{-+},\ \delta_4=\delta_{--}$$

to simplify our notation.

Let $C=C(H^2(M_f,\mathbb{Q}))$ (resp. $C^+=C^+(H^2(M_f,\mathbb{Q}))$) be the Clifford algebra (resp. the even Clifford algebra) of $(H^2(M_f,\mathbb{Q}), \psi_f)$ over K_f, and let us identify $H^2(M_f,\mathbb{Q})$ with a subspace of C by means of the canonical injection $H^2(M_f,\mathbb{Q}) \longrightarrow C$. The space $N=K_f\delta_1+K_f\delta_2$ is a maximal totally isotropic subspace of $H^2(M_f,\mathbb{Q})$ with respect to ψ_f. Therefore, as is known (cf. Bourbaki [7] for example), any element of C is naturally identified with an element of $\mathrm{End}_{K_f}(S)$. Here $S=\overset{*}{\Lambda} N$ is the exterior algebra of N over K_f, which is identified with a subalgebra of C naturally. Put $S_+=K_f+K_f\delta_1\wedge\delta_2$, and $S_-=K_f\delta_1+K_f\delta_2$. Then any element of C^+ preserves the subspaces S_+ and S_- of S. Counting the ranks of algebras, we have

$$C^+\cong \mathrm{End}_{K_f}(S_+)\oplus \mathrm{End}_{K_f}(S_-).$$

It is easy to check that the elements $e_1=\delta_1\delta_4\delta_2\delta_3+\delta_4\delta_1\delta_3\delta_2$ and $e_2=\delta_1\delta_4\delta_3\delta_2+\delta_4\delta_1\delta_2\delta_3$ of C^+ give the unity of $\mathrm{End}_{K_f}(S_+)$ and the unity of $\mathrm{End}_{K_f}(S_-)$, respectively. Therefore, e_1 and e_2 belong to the centre of C^+, and satisfy the relations:

$$e_1^2=e_1,\ e_2^2=e_2,\ e_1e_2=e_2e_1=0, \text{ and } e_1+e_2=1.$$

To find a simple factor (say) A_f^1 of the abelian variety $A(f)$, we choose an element $e_{11}=e_1(\delta_4\delta_1)=(\delta_4\delta_1)e_1$ of $C^+e_1\cong\mathrm{End}_{K_f}(S_+)$, which corresponds to the matrix

$$\begin{pmatrix}1&0\\0&0\end{pmatrix}$$

by means of the identification $\mathrm{End}_{K_f}(S_+)=M_2(K_f)$ with respect to the basis $\{1.\ \delta_1\wedge\delta_2\}$ of S_+. A simple calculation shows that

$$\delta_1\delta_3e_{11}=\delta_1\delta_4e_{11}=\delta_2\delta_3e_{11}=\delta_2\delta_4e_{11}=\delta_3\delta_4e_{11}=0,$$
$$1\cdot e_{11}=e_{11},\ \delta_1\delta_2e_{11}=\delta_1\delta_2,\ e_{11}e_{11}=e_{11}.$$

Therefore, we have

$$C^+e_{11} = K_f e_{11} + K_f \delta_1\delta_2,$$

as left C^+-modules.

Take the set $\{\sigma_1,\ldots,\sigma_d\}$ of all embeddings of K_f into $\mathbb{R}$. Then A_f^1 is given by

$$(\mathbb{R}^d e_{11} + \mathbb{R}^d \delta_1\delta_2)/L$$

as a real torus, where L is given by

$$L = \{ (\sigma_1(\lambda),\ldots,\sigma_d(\lambda))e_{11} + (\sigma_1(\mu),\ldots,\sigma_d(\mu))\delta_1\delta_2 \mid$$
$$\lambda \in m_f,\ \mu \in m_f' \text{ for some } \mathbb{Z}\text{-submodules } m_f \text{ and } m_f' \text{ of rank d}$$
$$\text{in } K_f \}.$$

Here we identify $K_f \otimes_{\mathbb{Q}} \mathbb{R}$ with $\mathbb{R}^d$ via the product of all embeddings σ_i.

Now let us see the action of the complex structure

$$J = \bigoplus_{i=1}^{d} J_i$$

defined in Section 5.7. It suffices to see the action of J_1 on

$$(C^+ \otimes_{K_f} \mathbb{R})e_{11} = \mathbb{R}e_{11} + \mathbb{R}\delta_1\delta_2,$$

for instance. In order to define J_1, we have to determine the intersection

$$H^2(M_f,\mathbb{Q}) \otimes_{K_f} \mathbb{R} \cap (H^{2,0} + H^{0,2})$$

in $H^2(M_f,\mathbb{Q}) \otimes_{K_f} \mathbb{C}$.

Let $\{\gamma_1, \gamma_2, \gamma_3, \gamma_4\}$ be the canonical basis of $H^2(M_f,\mathbb{Q})$ corresponding to $\{\delta_1, \delta_2, \delta_3, \delta_4\}$ by Poincaré duality. Moreover suppose that f is normalized, and put

$$W_{5-i} = \int_{\gamma_i} \omega_f \qquad (i=1,2,3,4).$$

Then in $H^2(M_f,\mathbb{Q}) \otimes_{K_f} \mathbb{C}$, ω_f is written as

$$\omega_f = W_1\delta_1 + W_2\delta_2 + W_3\delta_3 + W_4\delta_4,$$

and $\overline{\omega}_f$ as

$$\overline{\omega}_f = W_1\delta_1 - W_2\delta_2 + W_3\delta_3 - W_4\delta_4,$$

by Lemma 4.2. In view of the fact that $W_1, W_4 \in \mathbb{R}$ and $W_2, W_3 \in \sqrt{-1}\,\mathbb{R}$, we can readily show that

$$H^2(M_f,\mathbb{Q}) \otimes_{K_f} \mathbb{R} \cap (H^{2,0} + H^{0,2}) = \mathbb{R}(W_1\delta_1 + W_4\delta_4) + \mathbb{R}(\sqrt{-1}W_2\delta_2 + \sqrt{-1}W_3\delta_3)$$

Hence we can take as e_+ and e_-,

$$e_+=|W_1W_4|^{-1/2}(W_1\delta_1+W_4\delta_4) \quad \text{and} \quad e_-=|W_2W_3|^{-1/2}(\sqrt{-1}W_2\delta_2+\sqrt{-1}W_3\delta_3).$$

Then

$$J_1=e_+e_-=|W_1W_2W_3W_4|^{-1/2}\sqrt{-1}(W_1W_3\delta_1\delta_3+W_1W_2\delta_1\delta_2+W_2W_4\delta_4\delta_2+W_3W_4\delta_4\delta_3).$$

Thus we have

$$J_1e_{11}=|W_1W_2W_3W_4|^{-1/2}(\sqrt{-1}W_1W_2\delta_1\delta_2),$$
$$J_1\delta_1\delta_2=|W_1W_2W_3W_4|^{-1/2}(-\sqrt{-1}W_3W_4e_{11}).$$

Let

$$\kappa:\mathbb{R}e_{11}+\mathbb{R}\delta_1\delta_2 \longrightarrow \mathbb{C}$$

be the mapping, by means of which the J_1-multiplication corresponds to $\sqrt{-1}$-multiplication of $\mathbb{C}$. The mapping κ is written as

$$\kappa(ae_{11}+b\delta_1\delta_2)=a+bv\sqrt{-1}, \quad \text{for any} \quad a,\ b\in\mathbb{R},$$

with a real number v to be determined. Clearly $v\sqrt{-1}$ is given by

$$\begin{aligned} v\sqrt{-1}&=-|W_1W_2W_3W_4|^{-1/2}W_3W_4 & \\ &=-W_3W_4/|W_1W_4| & (|W_1W_4|=|W_2W_3|,\ \text{by Th. 4.4,(i)}) \\ &=-(W_3W_4)/(W_1W_4) & (W_1W_4>0,\ \text{by Th. 4.4, (ii)}) \\ &=W_4/W_2 & (W_1W_4+W_2W_3=0,\ \text{by Th. 4.4, (i)}) \\ &=W_{++}(f)/W_{-+}(f). & \end{aligned}$$

In view of other J_i all together, we can see that the period modulus of A_f^1 is given by the $GL_2(K_f)$-orbit of

$$(W_{-+}(f_i)/W_{++}(f_i))_{1\leqq i\leqq d} \qquad \text{in } X=(\mathbb{C}-\mathbb{R})^d.$$

It is easy to check that another choice of canobical basis of $H^2(M_f,\mathbb{Q})$ defines a point of X, which is $GL_2(K_f)$-equivalent to the above point. Replacing e_{11}, say, by $e_{21}=\delta_1\delta_2e_2$, we can discuss the case of $A_f^2\otimes\mathbb{Q}$ similarly. q.e.d.

§7. Main theorem A and its corollaries.

7.1. In order to formulate Main Theorem A and its corollaries, it is convenient to introduce some terminology.

Let K be a totally real algebraic number field. Then, a rational Hodge structure $H_{\mathbb{Q}}$ is called a K-Hodge structure (or a rational Hodge structure with coefficient in K in the terminology of Deligne [12], Section 2), if the field K acts on $H_{\mathbb{Q}}$ as endomorphisms of Hodge structure: $K \hookrightarrow \mathrm{End}(H_{\mathbb{Q}})$. In this case, for each embedding $\sigma: K \hookrightarrow \mathbb{C}$, the extension of scalars $H_{\mathbb{Q}} \otimes_{K,\sigma} \mathbb{C}$ naturally has a Hodge decomposition.

Let $H_{\mathbb{Q}}$ be a K-Hodge structure of weight n, then a K-bilinear form

$$\psi: H_{\mathbb{Q}} \times H_{\mathbb{Q}} \longrightarrow K$$

is called a K-polarization, if for any embedding $\sigma: K \hookrightarrow \mathbb{R}$, the extension of scalars $\psi \otimes_{K,\sigma} \mathbb{R}$ of ψ is a polarization of the real Hodge structure $H_{\mathbb{Q}} \otimes_{K,\sigma} \mathbb{R}$ of weight n.

We define the morphisms of K-Hodge structures in the evident manner. The category of K-Hodge structures is an abelian category.

7.2. <u>Main Theorem A</u>. <u>Let $H^2(M_f,\mathbb{Q})$ be the polarized rational Hodge structure attached to a primitive form f of $S_2(SL_2(O_F))$. And suppose that A_f^1 and A_f^2 are two abelian varieties constructed in Section 5.6 from the datum $\{ H^2(M_f,\mathbb{Q}), \theta_f, \psi_f \}$. Then we have an isomorphism of K_f-Hodge structures</u>

$$H^2(M_f,\mathbb{Q}) \cong H^1(A_f^1,\mathbb{Q}) \otimes_{K_f} H^1(A_f^2,\mathbb{Q}).$$

<u>Proof</u>. Let $\{\delta_{ss'} \mid s=\pm, s'=\pm\}$ be a canonical basis of $H^2(M_f,\mathbb{Q})$, and fix a set of normalized companions $\{ f_j \mid 1 \leqq j \leqq d \}$ as in Theorem 6.6.

Let $\{\sigma_1, \ldots, \sigma_d\}$ be the set of all embeddings of K_f into $\mathbb{C}$. Then we can choose a basis

$$\{ \omega_j^i \mid 1 \leqq j \leqq d \}$$

of the space of holomorphic 1-forms $\Gamma(A_f^i, \Omega^1_{A_f^i})$ over A_f^i (i=1,2), such that

$$\theta^{i*}(a)\omega_j^i = \sigma_j(a)\omega_j^i \quad \text{for any } a \quad K_f \text{ and each } j \ (1 \leqq j \leqq d),$$

where θ^i is the real multiplication

$$\theta^i: K_f \hookrightarrow \mathrm{End}(A_f^i).$$

Suppose that $\{\delta_+^i, \delta_-^i\}$ is a K_f-basis of $H^1(A_f^i,\mathbb{Q})$. For each i,j

$(i=1,\text{or } 2;\ 1\leqq j\leqq d)$, let $\delta^i_{+,j}$ and $\delta^i_{-,j}$ be the images of δ^i_+ and δ^i_- in $H^1(A^i_f,\mathbb{Q})\otimes_{K,\sigma_j}\mathbb{C}$, respectively. Then ω^i_j is written as

$$\omega^i_j = W^i_{+,j}\,\delta^i_{+,j} + W^i_{+,j}\,\delta^i_{+,j}$$

with some complex numbers $W^i_{+,j}$ and $W^i_{-,j}$ for each i,j.

Thanks to Theorem 6.6, we can choose the K_f-basis $\{\delta^i_+,\ \delta^i_-\}$ of $H^1(A^i_f,\ \mathbb{Q})$ such that

$$W^1_{+,j}/W^1_{-,j} = W_{+-}(f_j)/W_{++}(f_j)$$

and

$$W^2_{+,j}/W^2_{-,j} = W_{-+}(f_j)/W_{++}(f_j),$$

for each j $(1\leqq j\leqq d)$.

Let us define an isomorphism

$$\rho: H^2(M_f,\mathbb{Q}) \cong H^1(A^1_f,\mathbb{Q})\otimes_{K_f} H^1(A^2_f,\mathbb{Q})$$

of K_f-modules by the formulae

$$\rho(\delta_{++})=\delta^1_+\otimes_{K_f}\delta^2_+,\ \ \rho(\delta_{+-})=-\delta^1_+\otimes_{K_f}\delta^2_-,\ \ \rho(\delta_{-+})=-\delta^1_-\otimes_{K_f}\delta^2_+,$$

and

$$-\rho(\delta_{--})=\delta^1_-\otimes_{K_f}\delta^2_-.$$

Now let us check that this isomorphism ρ induces an isomorphism of K_f-Hodge structures. First, let us see the (2,0) type components of the Hodge structures.

Choose an embedding $\sigma=\sigma_j: K_f \hookrightarrow \mathbb{C}$, and consider the extension of scalars of ρ with respect to $\sigma=\sigma_j$:

$$\rho\otimes_{K_f,\sigma}\mathbb{C}:\ H^2(M_f,\mathbb{Q})\otimes_{K_f,\sigma}\mathbb{C} \cong \{H^1(A^1_f,\mathbb{Q})\otimes_{K_f} H^1(A^2_f,\mathbb{Q})\}\otimes_{K_f,\sigma}\mathbb{C}$$

$$= \{H^1(A^1_f,\mathbb{Q})\otimes_{K_f,\sigma}\mathbb{C}\}\otimes_{\mathbb{C}}\{H^1(A^2_f,\mathbb{Q})\otimes_{K_f,\sigma}\mathbb{C}\}.$$

The (2,0) component of $H^2(M_f,\mathbb{Q})\otimes_{K_f,\sigma}\mathbb{C}$ is given by 1-dimensional $\mathbb{C}$-linear space

$$\mathbb{C}\omega_{f_j} = \mathbb{C}\{W_{--}(f_j)\delta^\sigma_{++} + W_{-+}(f_j)\delta^\sigma_{+-} + W_{+-}(f_j)\delta^\sigma_{-+} + W_{++}(f_j)\delta^\sigma_{--}\}.$$

The (2,0) component of the right hand side of $\rho\otimes_{K_f,\sigma}\mathbb{C}$ is given by the tensor product of (1,0) components of $H^1(A^1_f,\mathbb{Q})\otimes_{K_f,\sigma}\mathbb{C}$ and $H^1(A^2_f,\mathbb{Q})\otimes_{K_f,\sigma}\mathbb{C}$. By the choice of $\{\delta^i_+,\ \delta^i_-\}$ $(i=1,2)$, the (1,0) component of $H^1(A^1_f,\mathbb{Q})\otimes_{K_f,\sigma}\mathbb{C}$ or $H^1(A^2_f,\mathbb{Q})\otimes_{K_f,\sigma}\mathbb{C}$ is given by

$$\mathbb{C}[\delta^1_{+,j} + \{W_{++}(f_j)/W_{+-}(f_j)\}\delta^1_{-,j}]$$

or

$$\mathbb{C}[\delta^2_{+,j} + \{W_{++}(f_j)/W_{-+}(f_j)\}\delta^2_{-,j}],$$

respectively.
Hence the (2,0) component of $\{H^1(A^1_f,\mathbb{Q})\otimes_{K_f}H^1(A^2_f,\mathbb{Q})\}\otimes_{K_f,\sigma}\mathbb{C}$ is given by

$$\mathbb{C}[\delta^1_{+,j}\otimes\delta^2_{-,j} + \{W_{++}(f_j)/W_{+-}(f_j)\}\delta^1_{-,j}\otimes\delta^2_{+,j}$$
$$+ \{W_{++}(f_j)/W_{-+}(f_j)\}\delta^1_{+,j}\otimes\delta^2_{-,j}$$
$$+ \{W_{++}(f_j)^2/W_{+-}(f_j)W_{-+}(f_j)\}\delta^1_{-,j}\otimes\delta^2_{-,j}].$$

In view of the period relation of Riemann-Hodge (Theorem 4.4)

$$W_{++}(f_j)W_{--}(f_j)+W_{+-}(f_j)W_{-+}(f_j)=0,$$

we can easily check that this space is equal to

$$(\rho\otimes_{K_f,\sigma}\mathbb{C})(\mathbb{C}\omega_{f_j}).$$

By Hodge symmetry, the (0,2) component of $H^2(M_f,\mathbb{Q})\otimes_{K_f,\sigma}\mathbb{C}$ is mapped to the (0,2) component of $\{H^1(A^1_f,\mathbb{Q})\otimes_{K_f}H^1(A^2_f,\mathbb{Q})\}\otimes_{K_f,\sigma}\mathbb{C}$. Accordingly, the (1,1) component is mapped to the (1,1) component by ρ. q.e.d.

7.3. Let $H_{\mathbb{Q}}$ be a rational Hodge structure of weight 2. Then by $b_2(H_{\mathbb{Q}})$ we denote the dimension of the vector space $H_{\mathbb{Q}}$ over $\mathbb{Q}$. The Picard number $\rho(H_{\mathbb{Q}})$ of $H_{\mathbb{Q}}$ is the dimension of the maximal rational Hodge substructure of $H_{\mathbb{Q}}$, which is purely of (1,1) type. We define the Lefschetz number $\lambda(H_{\mathbb{Q}})$ of $H_{\mathbb{Q}}$ by

$$\lambda(H_{\mathbb{Q}})=b_2(H_{\mathbb{Q}})-\rho(H_{\mathbb{Q}}).$$

Clearly, we have

$$\lambda(H_{\mathbb{Q}}+H'_{\mathbb{Q}}) = \lambda(H_{\mathbb{Q}}) + \lambda(H'_{\mathbb{Q}})$$

for any Hodge structures $H_{\mathbb{Q}}$ and $H'_{\mathbb{Q}}$ of weight 2. For the Hodge structure of Tate, $\lambda(\mathbb{Q}(-1))=0$.

By Lefschetz criterion on algebraic 2-cycles on algebraic surfaces, we have the following.

7.4. Scholium. (Lefschetz). *For the rational Hodge structure of the second cohomology group of a proper smooth connected surface over* $\mathbb{C}$,

<u>our definition of the Picard number and Lefschetz number coincide with the usual ones.</u>

<u>Remark</u>. For a Hilbert modular surface S, we have

$$\lambda(H^2_{sp}(S,\mathbb{Q})) = \lambda(W_2 H^2(S,\mathbb{Q})) = \lambda(H^2(S^*,\mathbb{Q})),$$

where S* is a smooth proper surface birationally equivalent to S (cf. Remark 1.13).

7.5. Let us denote by $H^2_{sp}(S,\mathbb{Q})_{alg}$ the subspace of $H^2_{sp}(S,\mathbb{Q})$ generated by the images of algebraic cycles. Since the action of Hecke operators holds algebraic cycles, $H^2_{sp}(S,\mathbb{Q})_{alg}$ is isomorphic to a direct sum of its direct factors

$$H^2(M_f,\mathbb{Q})_{alg} = H^2(M_f,\mathbb{Q}) \cap H^2_{sp}(S,\mathbb{Q})_{alg}.$$

Therefore, in order to know the Picard number of a Hilbert modular surface S, it suffices to know

$$H^2(M_f,\mathbb{Q})_{alg},\quad \rho(H^2(M_f,\mathbb{Q})),\quad \text{or } \lambda(M_f) \underset{dfn}{=} \lambda(H^2(M_f.\mathbb{Q}))$$

for primitive forms f of $S_2(SL_2(O_F))$.
Because of the trivial inequality

$$\rho \leqq h^{1,1}$$

between the Hodge number $h^{1,1}$ and the Picard number ρ, we have

$$2d \leqq \lambda(M_f) \leqq 4d.$$

Since $H^2(M_f,\mathbb{Q})_{alg}$ is a K_f-module, $\lambda(M_f)$ is a multiple of $d=[K_f:\mathbb{Q}]$:

$$\lambda(M_f)=2d,\ 3d,\ \text{or } 4d.$$

7.6. <u>Corollary 1 of Main Theorem A</u>. <u>Under the same notations as in Main Theorem A, we have an isomorphism of K_f-modules</u>

$$H^2(M_f,\mathbb{Q})_{alg} \cong \mathrm{Hom}_{O_f}(A^1_f,A^2_f)\otimes_{\mathbb{Z}}\mathbb{Q}.$$

<u>where O_f is a sufficiently small order of K_f such that</u>

$$\theta^i(O_f) \hookrightarrow \mathrm{End}(A^i_f) \qquad (i=1,2)$$

<u>for</u>

$$\theta^i : K_f \hookrightarrow \mathrm{End}(A^i_f)\otimes_{\mathbb{Z}}\mathbb{Q}.$$

<u>Proof</u>. By Main Theorem A, we have an isomorphism

$$H^2(M_f,\mathbb{Q}) \cong H^1(A^1_f,\mathbb{Q}) \otimes_{K_f} H^1(A^2_f,\mathbb{Q})$$

of K_f-Hodge structures. Since A^2_f is an abelian variety, there is a polarization

$$\Phi^2 : H^1(A_f^2,\mathbb{Q}) \times H^1(A_f^2,\mathbb{Q}) \longrightarrow \mathbb{Q}(-1).$$

Since A_f^2 is a Hilbert-Blumenthal abelain variety, $\Phi^2$is written as

$$\Phi^2 = \mathrm{tr}_{K_f/\mathbb{Q}}(\psi^2)$$

by a skew-symmetric K_f-bilinear form ψ^2 over $H^1(A_f^2,\mathbb{Q})$:

$$\psi^2 : H^1(A_f^2,\mathbb{Q}) \times H^1(A_f^2,\mathbb{Q}) \longrightarrow K_f.$$

Therefore, we have an isomorphism of K_f-Hodge structures

$$H^1(A_f^2,\mathbb{Q})^{\vee} = H^1(A_f^2,\mathbb{Q})(1),$$

where $H^1(A_f^2,\mathbb{Q})^{\vee}$ is the dual Hodge structure of $H^1(A_f^2,\mathbb{Q})$.
Hence we have an isomorphism of K_f-Hodge structures

$$H^2(M_f,\mathbb{Q}) \cong \underline{\mathrm{Hom}}_{K_f}(H^1(A_f^2,\mathbb{Q}),H^1(A_f^1,\mathbb{Q}))(-1),$$

where $\underline{\mathrm{Hom}}_{K_f}$ is the Hom-object in the category of rational K_f-Hodge structures.

By Lefschetz criterion on algebraic cycles over algebraic surfaces,

$$H^2(M_f,\mathbb{Q})_{alg} = H^2(M_f,\mathbb{Q}) \cap H^{1,1},$$

where $H^{1,1}$ is the (1,1) type component of $H^2(M_f,\mathbb{Q}) \otimes_{\mathbb{Q}} \mathbb{C}$.
Hence,

$$H^2(M_f,\mathbb{Q})_{alg} \cong \underline{\mathrm{Hom}}_{K_f}(H^1(A_f^2,\mathbb{Q}),\ H^1(A_f^2,\mathbb{Q})) \cap H^{0,0},$$

where $H^{0,0}$ is the (0,0) type component of $\underline{\mathrm{Hom}}_{K_f}(H^1(A_f^2,\mathbb{Q}),\ H^1(A_f^2,\mathbb{Q})) \otimes_{\mathbb{Q}} \mathbb{C}$.

Meanwhile the (0,0) component of

$$\underline{\mathrm{Hom}}_{K_f}(H^1(A_f^2,\mathbb{Q}),\ H^1(A_f^1,\mathbb{Q}))$$

is the homomorphisms $\mathrm{Hom}_{K_f}(H^1(A_f^2,\mathbb{Q}),\ H^1(A_f^1,\mathbb{Q}))$ of K_f-Hodge structures, whence our corollary follows. q.e.d.

7.7. Corollary 2 of Main Theorem A. Let f be a primitive form of $S_2(SL_2(O_F))$. Then we have the following equivalences:

(i) $\lambda(M_f)=4[K_f:\mathbb{Q}] \iff$ A_f and A_f are not K_f-isogenous.

(ii) $\lambda(M_f)=3[K_f:\mathbb{Q}] \iff$ A_f and A_f are K_f-isogenous, but neither is of CM-type.

(iii) $\lambda(M_f)=2[K_f:\mathbb{Q}] \iff$ There exists a CM-field $L \supset K_f$ with $[L:K_f]=2$, and with homomorphisms

$$\theta_L^i : L \hookrightarrow \mathrm{End}(A_f^i) \otimes_{\mathbb{Z}} \mathbb{Q} \quad (i=1,2),$$

extending the homomorphisms

$$\theta^i : K \hookrightarrow \mathrm{End}(A_f^i)\otimes_{\mathbb{Z}} \mathbb{Q}.$$

Proof. This corollary follows immediately from Corollary 1. In fact, $\lambda(M_f)=4[K_f:\mathbb{Q}]$ is equivalent to $H^2(M_f,\mathbb{Q})_{alg}=0$, which is equivalent to

$$\mathrm{Hom}_{O_f}(A_f^1, A_f^2)=0,$$

by Corollary 1. Hence the equivalence of (i) follows.

For (ii), by Corollary 1, $\lambda(M_f)=3[K_f:\mathbb{Q}]$ is equivalent to

$$\mathrm{rank}_{K_f}(\mathrm{Hom}_{O_f}(A_f^1,A_f^2)\otimes_{\mathbb{Z}}\mathbb{Q})=1.$$

Hence A_f^1 and A_f^2 are K_f-isogenous, and $\mathrm{End}_{O_f}(A_f^i)\otimes_{\mathbb{Z}}\mathbb{Q}$ is of rank 1 over K_f for each i.
Hence, A_f^i is not of CM-type. The converse is easy to check.

Let us check the last case (iii). By Corollary 1, $\lambda(M_f)=2[K_f:\mathbb{Q}]$ is equivalent to

$$\mathrm{rank}_{K_f}(\mathrm{Hom}_{O_f}(A_f^1,A_f^2)\otimes_{\mathbb{Z}}\mathbb{Q})=2.$$

This implies that A_f^1 and A_f^2 are K_f-isogenous, and that

$$\mathrm{rank}_{K_f}(\mathrm{End}_{O_f}(A_f^i)\otimes_{\mathbb{Z}}\mathbb{Q})=2.$$

By the structure theory of the endomorphism rings of abelian varieties (cf. Mumford [27] for example), we can check that the algebra

$$L=\mathrm{End}_{O_f}(A_f^1)\otimes_{\mathbb{Z}}\mathbb{Q}\cong\mathrm{End}_{O_f}(A_f^2)\otimes_{\mathbb{Z}}\mathbb{Q}$$

is a CM-field of degree 2 over K_f. The converse is also readily verified. q.e.d.

7.8. Remark. There exists a sorites to show Main Theorem A without explicit computation of the period moduli of A_f^i. In place of $H^2(M_f,\mathbb{Q})$, we start with a Hodge structure of weight 0

$$H^2(M_f,\mathbb{Q})^{\#}(1)\underset{\mathrm{dfn}}{=}H^2(M_f,\mathbb{Q})(1)\oplus\underline{K}_f,$$

where $\underline{K}_f$ is the trivial K_f-Hodge structure of rank 1 over K_f, and purely of (0,0) type. We can apply the formalism of Section 5. Then we have an abelian variety $A^{\#}$ of dimension $8[K_f:\mathbb{Q}]$. It is easy to check that $A^{\#}$ is isogenous to $A\times A\times A\times A$ with an abelian variety A which is isogenous to $A_f^1\times A_f^2$. By using Remark of Section 5.5, we can deduce Main Theorem A. Details are omitted.

§8. Selfconjugate forms and algebraic cycles.

Let us consider a mapping $\tilde{\iota}$ of $H \times H$ into $H \times H$ defined by

$$\tilde{\iota}:(z_1, z_2) \in H \times H \longmapsto (z_2, z_1) \in H \times H.$$

Since $\tilde{\iota}$ is contained in the normalizer of $SL_2(O_F)$ in the group of diffeomorphisms of $H \times H$, on passing to the quotient S, we have an involutive automorphism

$$\iota : S \longrightarrow S$$

of the surface S.

In this section and the next, we discuss some facts, derived from the existence of ι.

8.1. <u>Lemma</u>. $\iota \cdot G_\infty = H_\infty \cdot \iota$, <u>and</u> $\iota \cdot H_\infty = G_\infty \cdot \iota$.

<u>Proof</u>. A straightforword calculation shows the lemma. q.e.d.

It is easy to check that ι acts on $H^2(S,\mathbb{Q})$, $H_2(S,\mathbb{Q})$, $\tilde{H}_2(S,\mathbb{Q})$, $H_2(\partial S_M,\mathbb{Q})$, and $W_2H^2(S,\mathbb{Q})$, and moreover on $H_2^{sp}(S,\mathbb{Q})$ and $H^2_{sp}(S,\mathbb{Q})$, because

$$\iota^*(\eta_1) = \eta_2 \text{ and } \iota^*(\eta_2) = \eta_1.$$

It follows immediately from the above lemma that

$$\iota_*(H_2^{sp}(S,\mathbb{Q})_{++}) = H_2^{sp}(S,\mathbb{Q})_{++},$$
$$\iota_*(H_2^{sp}(S,\mathbb{Q})_{+-}) = H_2^{sp}(S,\mathbb{Q})_{-+},$$
$$\iota_*(H_2^{sp}(S,\mathbb{Q})_{-+}) = H_2^{sp}(S,\mathbb{Q})_{+-},$$
$$\iota_*(H_2^{sp}(S,\mathbb{Q})_{--}) = H_2^{sp}(S,\mathbb{Q})_{--}.$$

8.2. For any $f(z_1, z_2)$ of $S_2(SL_2(O_F))$, we write $\iota(f)(z_1, z_2)$ for $f(z_2, z_1)$. Then

$$\iota^*(\omega_f) = -\omega_{\iota(f)}.$$

For any ideal $\mathfrak{a}$ of O_F, we have a commutation relation

$$T_{\mathfrak{a}} \cdot \iota = \iota \cdot T_{\mathfrak{a}'}$$

of Hecke operators and ι. Here $\mathfrak{a}'$ is the conjugate ideal of $\mathfrak{a}$ over $\mathbb{Z}$. Therefore, especially, if f is a primitive form of $S_2(SL_2(O_F))$, then $\iota(f)$ is also a primitive form of $S_2(SL_2(O_F))$, and

$$K_f = K_{\iota(f)}.$$

8.3. Let H_0 be the $\mathbb{Q}$-subalgebra of $\mathrm{End}(H^2_{sp}(S,\mathbb{Q}))$ generated by the images of Hecke operators as in Section 2.1. Then, since

$$\iota T_{\mathfrak{a}}\iota = T_{\mathfrak{a}'} \quad \text{for any ideal } \mathfrak{a} \text{ of } O_F,$$

we have

$$\iota^* H_0 \iota^* = H_0 \quad \text{in } \mathrm{End}(H^2_{sp}(S,\mathbb{Q})).$$

Let us denote by ρ the involutive automorphism of H_0 defined by

$$\rho(a) = \iota^* a \iota^* \quad \text{for } a \in H_0.$$

Suppose that $\{e_1, \ldots, e_m\}$ be the complete system of the primitive idempotents of H_0. Then, $\{\rho(e_i) \mid 1 \leqq i \leqq m\}$ is also a complete system of primitive idempotents of H_0. Therefore $\rho(e_i) = e_j$ with some primitive idempotent e_j.

Three cases occur:

(i) $\rho(e_i) = e_i$, and ρ acts as identity on the field $e_i H_0 e_i$.

(ii) $\rho(e_i) = e_i$, and ρ induces a non-trivial automorphism of order 2 of the field $e_i H_0 e_i$.

(iii) $\rho(e_i) = e_j \neq e_i$.

By extension of scalars to $\mathbb{C}$, it is easy to check that

$$\rho(e_f) = e_{\iota(f)}$$

for any primitive form f of $S_2(SL_2(O_F))$.

From now on, in this section we discuss the case (i).

8.4. Definition. A primitive form f of $S_2(SL_2(O))$ is called selfconjugate, if $\iota(f)$ is a constant multiple of f.

If f is normalized, a selfconjugate form satisfies $f = \iota(f)$. Hence, $\rho(e_f) = e_{\iota(f)} = e_f$. Suppose that $T_{\mathfrak{a}} f = a_{\mathfrak{a}} f$ for each $\mathfrak{a}$ with $a_{\mathfrak{a}} \in K_f$. Then,

$$a_{\mathfrak{a}} f = T_{\mathfrak{a}} f = (\iota T_{\mathfrak{a}'} \iota) f = \iota(T_{\mathfrak{a}'} f) = \iota(a_{\mathfrak{a}'} f) = a_{\mathfrak{a}'} f.$$

Thus $a_{\mathfrak{a}} = a_{\mathfrak{a}'}$ for any ideal $\mathfrak{a}$. Via the identification $e_f H_0 e_f \cong K_f$, the action of ρ on $e_f H_0 e_f$ is transformed to the automorphism of K_f:

$$a_{\mathfrak{a}} \in K_f \longmapsto a_{\mathfrak{a}'} \in K_f.$$

Hence ρ acts trivially on $e_f H_0 e_f$. Therefore e_f is the case (i), if f is selfconjugate. Conversely, if e_f is the case (i), then

$$T_{\mathfrak{a}} f = T_{\mathfrak{a}'} f = (\iota T_{\mathfrak{a}} \iota) f \quad \text{for any ideal } \mathfrak{a} \text{ of } O_F.$$

Therefore f is a constant multiple of $\iota(f)$, by Multiplicity One Theorem (Theorem 2.2).

Remark. A selfconjugate form is usually called symmetric. We avoid this terminology, because for symmetric forms f, the corresponding holomorphic 2-forms ω_f are antisymmetric

$$\iota^*(\omega_f)=-\omega_f.$$

8.5. Lemma. If f is a selfconjugate form, then for any embedding $\sigma:K_f\hookrightarrow\mathbb{C}$ the companion f^σ is also selfconjugate.

Proof. This is clear, because the condition of selfconjugacy is equivalent to the coincidence of eigenvalues $a_{\mathfrak{a}}=a_{\mathfrak{a}'}$ for any ideal $\mathfrak{a}$. q.e.d.

Let us investigate the action of ι_* on $H_2(M_f,\mathbb{Q})_{\delta\delta'}$ $(\delta=\pm,\ \delta'=\pm)$. Each space of these four spaces is K_f-modules of rank 1. Note that the action of K_f and ι on $H_2(M_f,\mathbb{Q})$ or $H^2(M_f,\mathbb{Q})$ commute, since ρ acts on K_f trivially.

8.6. Lemma. Let f be a primitive form of $S_2(SL_2(\mathcal{O}_F))$. Let γ be an element of $H_2(M_f,\mathbb{Q})_{++}$ or $H_2(M,\mathbb{Q})_{--}$. Then, if

$$\int_\gamma \omega_{f^\sigma}=0$$

for any companion f^σ, the cycle $\gamma=0$.

Proof. In fact, $\gamma=G_\infty(\gamma)=H_\infty(\gamma)=F_\infty(\gamma)$, if $\gamma\in H_2(M_f,\mathbb{Q})_{++}$. Therefore,

$$\int_\gamma \omega_{f^\sigma}=\int_\gamma G_\infty^*(\omega_{f^\sigma})=\int_\gamma \eta_{f^\sigma,1}=\int_\gamma H_\infty^*(\omega_{f^\sigma})=\int_\gamma \eta_{f^\sigma,2}$$
$$=\int_\gamma F_\infty^*(\omega_{f^\sigma})=\int_\gamma \overline{\omega}_{f^\sigma}=0,$$

for any companion f^σ. Hence

$$\int_\gamma \omega=0,\quad \text{for any } \omega\in H^2(M_f,\mathbb{C}).$$

Therefore, $\gamma=0$. We can discuss similarly for $\gamma\in H_2(M_f,\mathbb{Q})_{--}$. q.e.d.

8.7. Proposition. Suppose that f is a selfconjugate form of $S_2(SL_2(\mathcal{O}_F))$. Then for any $\gamma\in H_2(M_f,\mathbb{Q})_{++}$ or $\gamma\in H_2(M_f,\mathbb{Q})_{--}$, we have

$$\iota_*(\gamma)=-\gamma.$$

Moreover

$$\iota_*(H_2(M_f,\mathbb{Q})_{+-})=H_2(M_f,\mathbb{Q})_{-+},$$

and

$$\iota_*(H_2(M_f,\mathbb{Q})_{-+})=H_2(M_f,\mathbb{Q})_{+-}.$$

Proof. Suppose that $\gamma\in H_2(M_f,\mathbb{Q})_{++}$ for instance, and put

$$\gamma_0=\gamma+\iota_*(\gamma).$$

Then $\gamma_0 \in H_2(M_f,\mathbb{Q})_{++}$, because $\iota_*(H_2(M_f,\mathbb{Q})_{++}) = H_2(M_f,\mathbb{Q})_{++}$ by Lemma 8.1. Since $\iota_*(\gamma_0)=\gamma_0$, we have

$$\int_{\gamma_0} \omega_f = \int_{\iota_*(\gamma_0)} \iota^*(\omega_f) = -\int_{\gamma_0} \omega_f = 0.$$

Therefore, by Lemma 8.6, we have $\gamma_0 = 0$, accordingly $\iota_*(\gamma) = -\gamma$. We can discuss the case $\gamma \in H_2(M_f,\mathbb{Q})_{--}$ similarly. The latter part of our proposition follows immediately from Lemma 8.1. q.e.d.

Now let us consider subspaces

$$H_2(M_f,\mathbb{Q})_-^{sym} = \{\ \gamma \in H_2(M_f,\mathbb{Q}) \mid F_\infty(\gamma) = -\gamma,\ \text{and}\ \iota_*(\gamma) = \gamma\ \}$$

and

$$H_2(M_f,\mathbb{Q})_-^{asym} = \{\gamma \in H_2(M_f,\mathbb{Q}) \mid F_\infty(\gamma) = -\gamma,\ \text{and}\ \iota_*(\gamma) = -\gamma\ \}$$

of $H_2(M_f,\mathbb{Q})_{+-} \oplus H_2(M_f,\mathbb{Q})_{-+}$. Then both spaces are K_f-modules of rank 1, because for any cycles $\gamma \in H_2(M_f,\mathbb{Q})_-^{sym}$ and $\gamma' \in H_2(M_f,\mathbb{Q})_-^{asym}$, we have

$$\psi_f(\gamma,\ \gamma') = \psi_f(\iota_*(\gamma),\ \iota_*(\gamma')) = -\psi_f(\gamma,\ \gamma') = 0,$$

and because of the non-degeneracy of ψ_f on the K_f-module of rank 2

$$H_2(M_f,\mathbb{Q})_{+-} \oplus H_2(M_f,\mathbb{Q})_{-+} = H_2(M_f,\mathbb{Q})_-^{sym} \oplus H_2(M_f,\mathbb{Q})_-^{asym}.$$

8.8. Proposition. *Let f be a selfconjugate form of $S_2(SL_2(O_F))$. Then, for any $g \in S_2(SL_2(O_F))$, and any $\gamma \in H_2(M_f,\mathbb{Q})_-^{sym}$, we have*

$$\int_\gamma \omega_g = 0.$$

Proof. Any element g of $S_2(SL_2(O_F))$ is written as a linear combination of primitive forms. Therefore, it suffices to show the statement of Proposition for primitive forms g. Let g be a primitive form, which is not a companion of f. Suppose that e_g and e_f are primitive idempotents of H_0, corresponding to g and f, respectively. Then,

$$\int_\gamma \omega_g = \int_{e_f(\gamma)} e_g \omega_g = \int_\gamma e_f e_g \omega_g = 0,$$

because $e_f e_g = 0$.

Let g be a companion of f. Then g is also selfconjugate by Lemma 8.5. Therefore,

$$\int_\gamma \omega_g = \int_{\iota_*(\gamma)} \omega_g = \int_\gamma \iota^*(\omega_g) = -\int_\gamma \omega_g = 0. \qquad \text{q.e.d.}$$

8.9. Remark. Let Ξ be a set of primitive forms of $S_2(SL_2(O_F))$ such that

$$H_2^{sp}(S,\mathbb{Q}) = \bigoplus_{f\in\Xi} H_2(M_f,\mathbb{Q}).$$

And let $\Xi^{s.c}$ be the subset of Ξ consisting of all selfconjugate elements in Ξ. Then by Proposition 8.8, the cycles in

$$\bigoplus_{f\in\Xi^{s.c}} H_2(M_f,\mathbb{Q})_{-}^{sym}$$

are algebraic cycles by the criterion of Lefschetz. The dimension of of this space of algebraic cycles is equal to the dimension of selfconjugate forms. These facts are shown by Hirzebruch [17] in a slightly different formulation:in [17] symmetric Hilbert modular surfaces are discussed.

By Proposition 8.7, for selfconjugate forms f, we have

$$H_2(M_f,\mathbb{Q})^{sym} \underset{dfn}{=} \{\gamma\in H_2(M_f,\mathbb{Q}) \mid \iota_*(\gamma)=\gamma\} = H_2(M_f,\mathbb{Q})_{-}^{sym}.$$

Reformulating the results of Sections 8.7, 8.8, and 8.9 in terms of cohomology groups, we have the following theorem for selfconjugate forms.

8.10. Proposition. If f is selfconjugate, then we have a decomposition of K_f-Hodge structures

$$H^2(M_f,\mathbb{Q}) = H^2(M_f,\mathbb{Q})^{asym} \oplus H^2(M_f,\mathbb{Q})^{sym}.$$

Here

$$H^2(M_f,\mathbb{Q})^{sym} = \{\delta\in H^2(M_f,\mathbb{Q}) \mid \iota^*(\delta)=\delta\}$$

is a K_f-Hodge structure of rank 1 over K_f, which is purely of type (1,1), and here

$$H^2(M_f,\mathbb{Q})^{asym} = \{\delta\in H^2(M_f,\mathbb{Q}) \mid \iota^*(\delta)=-\delta\}$$

is K_f-Hodge structure of rank 3, such that for any embedding $\sigma:K_f\hookrightarrow\mathbb{C}$

$$H^2(M_f,\mathbb{Q})^{asym}\otimes_{K_f,\sigma}\mathbb{C}$$

has Hodge number $h^{2,0}=h^{1,1}=h^{0,2}=1$.
The space $H^2(M_f,\mathbb{Q})^{sym}$ is generated by the Poincaré duals of algebraic cycles.

By this proposition and Corollary 2 of Main Theorem A (Section 7.7), the abelian varieties A_f^1 and A_f^2 are K_f-isogenous, if the primitive form f of $S_2(SL_2(O_F))$ is selfconjugate. More precisely we have the following result.

8.11. Proposition. Let f be a selfconjugate primitive form of $S_2(SL_2(O_F))$. Then there exists an element a ($\neq 0$) of K_f such that

$$W_{-+}(f_i)=\sigma_i(a)W_{+-}(f_i) \quad \text{for all } i \ (1 \leqq i \leqq d),$$

where $\{\sigma_1,\ldots,\sigma_d\}$ is the set of all embeddings of K_f into $\mathbb{C}$, and each f_i is a companion of f with respect to σ_i. Here the periods $W_{+-}(f_i)$ and $W_{-+}(f_i)$ are defined as in Theorem 4.4 with respect to some canonical basis of $H_2(M_f,\mathbb{Q})$.
Especially A_f^1 and A_f^2 are K_f-isogenous, if f is selfconjugate.
Proof. Let

$$\{ \gamma_{\delta\delta'} \mid \delta=\pm,\ \delta'=\pm \}$$

be a canonical basis of $H_2(M_f,\mathbb{Q})$. Then γ_{+-} or γ_{-+} spans $H_2(M_f,\mathbb{Q})_{+-}$ or $H_2(M_f,\mathbb{Q})_{-+}$ over K_f, respectively. On the other hand

$$\iota_*(H_2(M_f,\mathbb{Q})_{-+})=H_2(M_f,\mathbb{Q})_{+-}.$$

Therefore,

$$\iota_*(\gamma_{-+})=\theta(a)\gamma_{+-}$$

for some $a\in K_f$.
Since the form f is selfconjugate,

$$\iota^*(\omega_f)=-\omega_f.$$

Hence our proposition follows.

The latter part follows immediately from Theorems 6.5 and 6.6.
q.e.d.

§9. Abelian varieties attached to non-selfconjugate forms.

In this section, we discuss the Hodge structures $H^2(M_f,\mathbb{Q})$ and the abelian varieties A_f^i $(i=1,2)$ attached to non-selfconjugate primitive forms f of $S_2(SL_2(O_F))$. We discuss only the case (ii) of Section 8.3, because the case (iii) does not occur in fact.

9.1. Assume that a primitive form f of $S_2(SL_2(O_F))$ is not selfconjugate. Then the corresponding primitive idempotent e_f of H_0 belongs to the case (ii) or (iii) in Section 8.3, according as

$$\rho(e_f)=e_{\iota(f)}=e_f \quad \text{or} \quad \rho(e_f)=e_{\iota(f)}\neq e_f.$$

In either case, by restriction, the automorphism ρ of H_0 induces an isomorphism of fields

$$e_f H_0 e_f \xrightarrow{\sim} e_{\iota(f)} H_0 e_{\iota(f)},$$

which is also denoted by ρ.

Meanwhile, restricted to direct factors, the automorphism ι^* of $H^2_{sp}(S,\mathbb{Q})$ induces an isomorphism of Hodge structures

$$H^2(M_f,\mathbb{Q}) \xrightarrow{\sim} H^2(M_{\iota(f)},\mathbb{Q}),$$

which is also denoted by ι^*.

Let ρ_f be the composition

$$K_f \simeq e_f H_0 e_f \xrightarrow[\rho]{\sim} e_{\iota(f)} H_0 e_{\iota(f)} \simeq K_{\iota(f)}$$

of the isomorphisms of fields. Then by the definitions, we find that $\iota^*: H^2(M_f,\mathbb{Q}) \xrightarrow{\sim} H^2(M_{\iota(f)},\mathbb{Q})$ is compatible with the actions of K_f and $K_{\iota(f)}$ via θ_f and $\theta_{\iota(f)}$:

$$\theta_{\iota(f)}(\rho_f(a))\iota^*(\delta)=\iota^*(\theta_f(a)\ )$$

for any $a\in K_f$ and any $\delta\in H^2(M_f,\mathbb{Q})$.

Because $K_f=K_{\iota(f)}$ as subfields of $\mathbb{R}$, ρ_f defines an automorphism of K_f, of order at most 2. It is easy to see that the action of ρ_f on K_f is given by

$$\rho_f(a_{\mathfrak{a}})=a_{\mathfrak{a}'}$$

for the eigenvalues $a_{\mathfrak{a}}: T_{\mathfrak{a}}f=a_{\mathfrak{a}}f$ of Hecke operators $T_{\mathfrak{a}}$. Since f is not selfconjugate, $a_{\mathfrak{a}}\neq a_{\mathfrak{a}'}$ for some ideal $\mathfrak{a}$ of O_F by Multiplicity One Theorem (Theorem 2.2). Hence ρ_f is not the identity of K_f. Hence the subfield k_f of K_f defined by

$$k_f=\{\ a\in K_f \mid \rho_f(a)=a\ \}$$

is of degree 2 in K_f, i.e. $[K_f:k_f]=2$.
Since K_f is a totally real algebraic number field, its subfield k_f is also totally real.

Now note that the composition of ρ_f with the canonical embedding $K_f \hookrightarrow \mathbb{C}$ defines another embedding of K_f into $\mathbb{C}$. Let f^ρ be a companion of f with respect to this embedding. Then two primitive forms $\iota(f)$ and f^ρ have the same eigenvalues with respect to the Hecke operators. Therefore, by Multiplicity One Theorem, $\iota(f)$ is a constant multiple of f^ρ. Hence

$$e_{\iota(f)}=e_{f^\rho}=e_f.$$

Therefore the case (iii) of Section 8.3 does not occur.

9.2. Because for any non-selfconjugate primitive form f of $S_2(SL_2(O_F))$ we always have

$$e_f=e_{\iota(f)},$$

the direct factors $H^2(M_f,\mathbb{Q})$ and $H^2(M_{\iota(f)},\mathbb{Q})$ of $H^2_{sp}(S,\mathbb{Q})$ coincide in $H^2_{sp}(S,\mathbb{Q})$. Therefore

$$\iota^*:H^2(M_f,\mathbb{Q}) \longrightarrow H^2(M_{\iota(f)},\mathbb{Q}) = H^2(M_f,\mathbb{Q})$$

is a ρ_f-linear automorphism of the K_f-Hodge structure $H^2(M_f,\mathbb{Q})$.

Define a new K_f-Hodge structure $H^2(M_f,\mathbb{Q})^\rho$ by

$$H^2(M_f,\mathbb{Q})^\rho= H^2(M_f,\mathbb{Q})\otimes_{K_f,\rho_f} K_f.$$

Namely, $H^2(M_f,\mathbb{Q})^\rho$ is identical with $H^2(M_f,\mathbb{Q})$ as a Hodge structure, but has a new action of K_f given by the composition

$$\theta_f^\rho:K_f \xrightarrow[\rho_f]{\sim} K_f \xhookrightarrow[\theta_f]{} \mathrm{End}(H^2(M_f,\mathbb{Q})).$$

Then ι^* naturally defines an isomorphism of K_f-Hodge structure

$$\nu:H^2(M_f,\mathbb{Q}) \xrightarrow{\sim} H^2(M_f,\mathbb{Q})^\rho,$$

which is the same homomorphism as ι^* as homomorphisms of rational Hodge structures.

Suppose that $C^+(H^2(M_f,\mathbb{Q}))$ and $C^+(H^2(M_f,\mathbb{Q})^\rho)$ be the even Clifford algebras with respect to ψ_f over $H^2(M_f,\mathbb{Q})$ and $H^2(M_f,\mathbb{Q})^\rho$, respectively. And denote by the same symbole ν the isomorphism of $C^+(H^2(M_f,\mathbb{Q}))$ to $C^+(H^2(M_f,\mathbb{Q})^\rho)$ induced from ν.

Let $\{ \delta_{ss'} \mid s=\pm, s'=\pm\}$ be a canonical basis of $H^2(M_f,\mathbb{Q})$. Then this basis gives also a canonical basis of $H^2(M_f,\mathbb{Q})$. Recall that the elements

$$e_1 = \delta_{++}\delta_{--}\delta_{+-}\delta_{-+} + \delta_{--}\delta_{++}\delta_{-+}\delta_{+-}$$

and

$$e_2 = \delta_{++}\delta_{--}\delta_{-+}\delta_{+-} + \delta_{--}\delta_{++}\delta_{+-}\delta_{-+}$$

are idempotents of $C^+(H^2(M_f,\mathbb{Q}))$ or $C^+(H^2(M_f,\mathbb{Q})^\rho)$, which correspond to the identities of the factors $M_2(K_f)$ via the isomorphism

$$C^+(H^2(M_f,\mathbb{Q})) = C^+(H^2(M_f,\mathbb{Q})^\rho) \cong M_2(K_f) \oplus M_2(K_f)$$

(cf. Section 5.6 and the proof of Theorem 6.6).
The images of e_1 and e_2 by the isomorphism ν are given by

$$\begin{aligned}\nu(e_1) &= \nu(\delta_{++}\delta_{--}\delta_{+-}\delta_{-+}) + \nu(\delta_{--}\delta_{++}\delta_{-+}\delta_{+-}) \\ &= \delta_{++}\delta_{--}\delta_{-+}\delta_{+-} + \delta_{--}\delta_{++}\delta_{+-}\delta_{-+} \\ &= e_2\end{aligned}$$

and

$$\nu(e_2) = e_1.$$

Let $A(f)$ and $A(f)^\rho$ be the abelian varieties constructed from the Clifford algebras $C^+(H^2(M_f,\mathbb{Q}))$ and $C^+(H^2(M_f,\mathbb{Q})^\rho)$ as in Section 5.6, by choosing some lattices in $H^2(M_f,\mathbb{Q})$ and $H^2(M_f,\mathbb{Q})^\rho$, respectively. Then, the isomorphism

$$\nu : H^2(M_f,\mathbb{Q}) \xrightarrow{\sim} H^2(M_f,\mathbb{Q})^\rho$$

of the Hodge structures induces an isogeny

$$\nu_A : A(f) \longrightarrow A(f)^\rho,$$

which satisfies

$$\nu_A(\Theta(\beta)x) = \Theta^\rho(\nu(\beta))\nu_A(x)$$

for any $\beta \in C^+ = C^+(H^2(M_f,\mathbb{Q}))$ and $x \in A(f)$, where

$$\Theta : C^+(H^2(M_f,\mathbb{Q})) \hookrightarrow \mathrm{End}(A(f))$$

and

$$\Theta^\rho : C^+(H^2(M_f,\mathbb{Q})^\rho) \hookrightarrow \mathrm{End}(A(f)^\rho)$$

are the natural homomorphisms induced from the right multiplication of $C^+(H^2(M_f,\mathbb{Q}))$ and $C^+(H^2(M_f,\mathbb{Q})^\rho)$ on themselves.

As we have seen in Section 5.6, $A(f)$ is isogenous to a product $A_f^1 \times A_f^1 \times A_f^2 \times A_f^2$ of Hilbert-Blumenthal abelian varieties A_f^1 and A_f^2 with homomorphisms

$$\theta_i : K_f \hookrightarrow \mathrm{End}(A_f^i)\otimes_{\mathbb{Z}}\mathbb{Q} \quad (i=1,2).$$

We define new Hilbert-Blumenthal abelian varieties $(A_f^{i\rho}, \theta_i^{\rho})$ $(i=1,2)$ from (A_f^i, θ_i), putting

$$A_f^{i\rho} = A_f^i$$

as abelian varieties and replacing θ_i by the compositions

$$\theta_i^{\rho} = \theta_i \circ \rho_f : K_f \xrightarrow{\sim} K_f \hookrightarrow \mathrm{End}(A_f^i)\otimes_{\mathbb{Z}}\mathbb{Q}.$$

Then $A(f)^{\rho}$ is isogenous to a product

$$A_f^{1\rho} \times A_f^{1\rho} \times A_f^{2\rho} \times A_f^{2\rho}.$$

9.3. Theorem. Let f be a non-selfconjugate primitive form of $S_2(SL_2(O_F))$. Then, A_f^1 and $A_f^{2\rho}$ are K_f-isogenous, or equivalently A_f^2 and $A_f^{1\rho}$ are K_f-isogenous.

Proof. Since $\nu(e_1)=e_2$ and $\nu(e_2)=e_1$, the isogeny

$$\nu_A : A(f) \longrightarrow A(f)^{\rho}$$

defined in Section 9.2 induces K_f-isogenies

$$\nu_A^1 : A_f^1 \longrightarrow A_f^{2\rho}$$

and

$$\nu_A^2 : A_f^2 \longrightarrow A_f^{1\rho}$$

of Hilbert-Blumenthal abelian varieties with respect to K_f, by restricting it to the direct factors of $A(f)$. q.e.d.

9.4. Corollary of Theorem 9.3. If f is a non-selfconjugate primitive form of $S_2(SL_2(O_F))$, and if the Lefschetz number $\lambda(M_f) = 3[K_f:\mathbb{Q}]$. Then the abelian varieties A_f^1 and A_f^2 are K_f-isogenous, and there exists a totally indefinite quaternion algebra B over k_f with an injective algebra homomorphism $K_f \longrightarrow B$ over k_f such that

$$B \hookrightarrow \mathrm{End}(A_f^1)\otimes_{\mathbb{Z}}\mathbb{Q} \cong \mathrm{End}(A_f^2)\otimes_{\mathbb{Z}}\mathbb{Q}.$$

Proof. There exists a K_f-isogeny

$$\alpha : A_f^1 \longrightarrow A_f^2$$

by Corollary 2 of Main Theorem A (Section 7.7). And by Theorem 9.3, there exists a K_f-isogeny

$$\nu_A : A_f^2 \longrightarrow A_f^{1\rho}.$$

Composing these two K_f-isogenies, we have a K_f-isogeny

$$\beta=\nu_A\circ\alpha:A_f^1 \longrightarrow A_f^{1\rho}.$$

Let B be the subalgebra of $End(A_f^1)\otimes_{\mathbb{Z}}\mathbb{Q}$ generated by β over $\theta_1(K_f)$. Since $\theta_1(a)\beta=\beta\theta_1(\rho_f(a))$ for any $a\in K_f$, B is a non-commutative algebra over k_f via θ_1 such that

$$\dim_{k_f} B= 2\cdot\dim_{K_f}\theta_1(K_f)[\beta]\geqq 4.$$

Since ρ_f is of order 2, $\beta\circ\beta=\beta^2:A_f^1 \longrightarrow A_f^{1\rho} \longrightarrow A_f^{1\rho\rho}=A_f^1$ commutes with the elements of $\theta_1(K_f)$. Because Corollary 2 of Theorem A implies

$$\mathrm{rank}_{K_f}\mathrm{End}_{O_f}(A_f^i)\otimes_{\mathbb{Z}}\mathbb{Q}=1$$

in our case, β^2 is an element of $\theta_1(K_f)$. Thus $\dim_{K_f}\theta_1(K_f)[\beta]\leqq 4$. Hence, $\dim_{k_f} B=4$.

Therefore B is a quaternion algebra over a totally real field k_f. Since it splits over a totally real extension K_f of k_f of degree 2, B is totally indefinite. q.e.d.

9.5. <u>Corollary of Theorem</u> 9.3. <u>If</u> f <u>is a non-selfconjugate primitive forms of</u> $S_2(SL_2(O_F))$ <u>with</u> $\lambda(M_f)=2[K_f:\mathbb{Q}]$. <u>Then</u> A_f^1 <u>and</u> A_f^2 <u>are</u> K_f<u>-isogenous and of CM-type with a CM-field</u> L_f <u>which is a composite of the quadratic extension</u> K_f/k_f <u>with a quadratic extension</u> S_f/k_f. <u>Here</u> S_f <u>is also a CM-field</u>.
<u>Moreover</u> A_f^1 <u>and</u> A_f^2 <u>are not simple, but isogenous to an isotypic product of a simple abelian variety</u>.
<u>Proof</u>. Since $\lambda(M_f)<3[K_f:\mathbb{Q}]$, applying the argument of the proof of Corollary of Section 9.4 , we can show that there exists a K_f-isogeny

$$\beta:A_f^1 \longrightarrow A_f^{1\rho}.$$

By Corollary 2 of Main Theorem A (Section 7.7), the algebra

$$L_f=\mathrm{End}_{O_f}(A_f^1)\otimes_{\mathbb{Z}}\mathbb{Q}\cong \mathrm{End}_{O_f}(A_f^2)\otimes_{\mathbb{Z}}\mathbb{Q}$$

is a CM-field.

Let us define an automorphism ρ_0 of $End(A_f^1)\otimes_{\mathbb{Z}}\mathbb{Q}$ by

$$\rho_0:\alpha \longmapsto \beta^{-1}\alpha\beta \quad \text{for any } \alpha\in End(A_f^1)\otimes_{\mathbb{Z}}\mathbb{Q}.$$

Then it is easy to check that, if α commutes with any element of $\theta_1(K_f)$, so does $\rho_0(\alpha)$. In fact, if

$$\alpha\theta_1(a)=\theta_1(a)\alpha \quad \text{for any } a\in K_f,$$

then

$$\beta^{-1}\alpha\beta\cdot\beta^{-1}\theta_1(a)\beta=\beta^{-1}\theta_1(a)\beta\cdot\beta^{-1}\alpha\beta,$$

which means

$$\rho_0(\alpha)\theta_1(a)=\theta_1(a)\rho_0(\alpha) \quad \text{for any } a\in K_f.$$

Thus ρ_0 induces an automorphism of $L_f=\mathrm{End}_{0_f}(A_f^1)\otimes_{\mathbb{Z}}\mathbb{Q}$, where 0_f is a sufficiently small order in K_f as in Section 7.7. Since ρ_0 induces ρ_f on K_f, ρ_0 is a non-trivial automorphism of order 2 of L_f. Let us denote this subfield by S_f. Then S_f/k_f is a quadratic extension, and $S_f\neq K_f$ in L_f. Since $[L_f:k_f]=[L_f:K_f][K_f:k_f]=4$, L_f is a composite of S_f and K_f. K_f is a totally real extension of k_f. Therefore, since L_f/K_f is a totally imaginary quadratic extension of a totally real field, so is S_f/k_f.

Let

$$A_f^1 \sim A_1\times A_2\times \ldots \times A_s$$

be a decomposition of A_f^1 into a product of products A_i^1 of isotypic factors up to isogeny. Then

$$B \hookrightarrow \mathrm{End}(A_f^1)\otimes_{\mathbb{Z}}\mathbb{Q}\cong \bigoplus_{j=1}^{s} \mathrm{End}(A_j)\otimes_{\mathbb{Z}}\mathbb{Q}.$$

Since A_f^1 is a Hilbert-Blumenthal abelian variety with respect to K_f ($\subset B$) and B is a simple algebra, s=1. Therefore A_f^1 is isogenous to a product

$$A\times \ldots \times A$$

of copies of a simple abelian variety A of CM-type. q.e.d.

9.6. Let us discuss here a relation of non-selfconjugate forms and the Hodge structures of the second cohomology groups of symmetric Hilbert modular surfaces.

Assume that f is not selfconjugate, and put

$$H^2(M_f^+,\mathbb{Q}) \underset{\mathrm{dfn}}{=} \{ \gamma\in H^2(M_f,\mathbb{Q}) \mid \iota(\gamma)=\gamma \}.$$

Then $H^2(M_f^+,\mathbb{Q})$ is a k_f-module of rank 4. We define the rational Hodge structure of weight 2 $H^2(M_f^+,\mathbb{Q})$ similarly.

Let S^+ be the symmetric Hilbert modular surface

$$S^+ \underset{\mathrm{dfn}}{=} \langle SL_2(0_F),\iota\rangle\backslash(H\times H).$$

Then there is a natural monomorphism of Hodge structures

$$H^2(M_f^+,\mathbb{Q}) \hookrightarrow W_2H^2(S^+,\mathbb{Q})= (W_2H^2(S,\mathbb{Q}))^+.$$

It is clear that there is an isomorphism of rational Hodge structures

$$H^2(M_f^+,\mathbb{Q})\otimes_{k_f}K_f \cong H^2(M_f,\mathbb{Q})$$

for any non-selfconjugate forms f. And moreover we can find a symmetric k_f-bilinear form

$$\psi_f^+ : H^2(M_f^+,\mathbb{Q}) \times H^2(M_f^+,\mathbb{Q}) \longrightarrow k_f,$$

such that

$$\psi_f^+\otimes_{k_f}K_f = a\psi_f \quad \text{for some } a\in K_f.$$

By using the decomposition

$$H^2(M_f,\mathbb{Q}) = \bigoplus_{s,s'} H^2(M_f,\mathbb{Q})_{ss'}$$

and noting the fact that

$$\iota^*(H^2(M_f,\mathbb{Q})_{++}) = H^2(M_f,\mathbb{Q})_{++}, \quad \iota^*(H^2(M_f,\mathbb{Q})_{--}) = H^2(M_f,\mathbb{Q})_{--},$$

$$\iota^*(H^2(M_f,\mathbb{Q})_{+-}) = H^2(M_f,\mathbb{Q})_{-+}, \quad \iota^*(H^2(M_f,\mathbb{Q})_{-+}) = H^2(M_f,\mathbb{Q})_{+-},$$

we can check easily that the quadratic form ψ_f^+ is chosen so that it is a direct sum of a kenel form of two variables (represented by a matrix $\left(\begin{smallmatrix}1 & 0\\ 0 & 1\end{smallmatrix}\right)$) and a norm form N_{K_f/k_f} of the quadratic extension K_f/k_f. Therefore the determinant of the quadratic form ψ_f^+ over k_f is equal to the discriminant of the extension K_f/k_f mod $(k_f^\times)^2$.

Let us apply the formalism of Section 5 to this k_f-polarized Hodge structure $(H^2(M_f^+,\mathbb{Q}),\ \psi_f^+)$. Then we obtain an abelian variety $A(f)^+$ of dimension $4[k_f:\mathbb{Q}] = 2[K_f:\mathbb{Q}]$. The Clifford algebra

$$C^+(H^2(M_f^+,\mathbb{Q}))$$

is isomorphic to $M_2(K_f)$ over k_f, because the discriminant of ψ_f^+ is the discriminant of K_f/k_f. Therefore $A(f)^+$ is isogenous to a product

$$A_f^+\times A_f^+$$

of an abelian variety A_f^+ of dimension $[K_f:\mathbb{Q}]$ over $\mathbb{C}$ with an endomorphism

$$K_f \hookrightarrow \mathrm{End}(A_f^+)\otimes_{\mathbb{Z}}\mathbb{Q}.$$

The above abelian variety A_f^+ is K_f-isogenous to both A_f^1 and A_f^2, corresponding to the two embeddings of K_f into K_f over k_f.

Since we do not use these facts later, details and proofs are omitted.

Chapter III. Correspondence betweeen real Nebentype elliptic modular forms and Hilbert modular forms.

Throughout this chapter, we denote by G the algebraic group $Res_{F/\mathbb{Q}}SL_2$ defined over $\mathbb{Q}$, where Res is the Weil's restriction of scalar functor. Moreover we identify its integral points $G_{\mathbb{Z}}$ (res. rational points $G_{\mathbb{Q}}$, resp. real points $G_{\mathbb{R}}$) with $SL_2(O_F)$ (resp. $SL_2(F)$, resp. $SL_2(\mathbb{R})\times SL_2(\mathbb{R})$).

In this chapter, we discuss the correspondence between real Nebentype elliptic modular forms and Hilbert modular forms via theta series attached to a quadratic form of four variables with signature (2+,2-).

§10. Weil representation and theta series.

10.1. In the first place, we define a metric lattice L of rank 4 over $\mathbb{Z}$ with $G_{\mathbb{Z}}$ action, given by

$$L=\{\ \ell\in M_2(F)\ |\ \ell=\begin{pmatrix} a & \lambda' \\ \lambda & b\end{pmatrix},\ \text{with}\ a,\ b\in\mathbb{Z},\ \text{and}\ \lambda\in\delta_F^{-1}\ \},$$

where δ_F^{-1} is the codifferent of F.

The metric on L is given by a symmetric bilinear form B and a quadratic form Q on L defined by

$$B(\ell_1,\ell_2)=D\cdot tr\{\begin{pmatrix} 0 & 1 \\ -1 & 0\end{pmatrix}{}^t\ell_1\begin{pmatrix} 0 & 1 \\ -1 & 0\end{pmatrix}\ell_2\}$$

$$=-D(a_1b_2+b_1a_2)+D(\lambda_1\lambda_2'+\lambda_2\lambda_1'),$$

and

$$Q(\ell_1)=B(\ell_1,\ell_1)=-2D(a_1b_1-\lambda_1\lambda_1'),$$

for $\ell_1=\begin{pmatrix} a_1 & \lambda_1' \\ \lambda_1 & b_1\end{pmatrix}$, $\ell_2=\begin{pmatrix} a_2 & \lambda_2' \\ \lambda_2 & b_2\end{pmatrix}\in L$.

Then Q is an even integral quadratic form on L.

Let us define an action of $G_{\mathbb{Z}}$ on L (or an action of $G_{\mathbb{Q}}$ on $L_{\mathbb{Q}}=L\otimes_{\mathbb{Z}}\mathbb{Q}$) by

$$\gamma(\ell)\underset{dfn}{=}\gamma\cdot\ell\cdot{}^t\gamma',\quad \text{for}\ \gamma\in G_{\mathbb{Z}}\ (\text{or}\ \gamma\in G_{\mathbb{Q}})\ \text{and}\ \ell\in L\ (\text{or}\ \ell\in L_{\mathbb{Q}}).$$

Here the right hand side is the multiplication of the matrices γ, ℓ, and ${}^t\gamma'$.

By extension of scalars, we can define an action of $G_{\mathbb{R}}$ on $L_{\mathbb{R}}=L\otimes_{\mathbb{Z}}\mathbb{R}$. Then the bilinear form B and the associated quadratic form Q, and their extensions of scalars are invariant under $G_{\mathbb{Z}}$, $G_{\mathbb{Q}}$, and $G_{\mathbb{R}}$.

Let L* be the dual lattice of L. Then, by means of B, we can identify L* with

$$L^* = \{\ \ell^* \in M_2(F)\ |\ \ell^{*\prime} = {}^t\ell^*,\ D\ell^* \in M_2(O_F)\ \}.$$

Clearly the index $[L^*;L]=D^3$, and the index $[L:DL^*]=D$. Therefore the level and the discriminant of the even integral quadratic form Q is D and D^3, respectively.

Let us consider a domain X in $L\otimes_{\mathbb{Z}}\mathbb{C}$ defined by

$$X = \{\ v = \begin{pmatrix} v_1 & v_2 \\ v_3 & v_4 \end{pmatrix} \in L\otimes_{\mathbb{Z}}\mathbb{C}\ |\ B(v,v)=0,\ B(v,\bar{v})<0\ \},$$

where $\bar{v}$ is the complex conjugate of v.
Let us consider a mapping $\mu: H\times H \longrightarrow X$ defined by

$$\mu((z_1,\ z_2)) = \begin{pmatrix} z_1z_2 & z_1 \\ z_2 & 1 \end{pmatrix}.$$

Let X^+ be one of the two connected componets of X, which contains the image of μ. Then, X^+ is evidently isomorphic to $H\times H\times \mathbb{C}^\times$.

For any $\gamma=\begin{pmatrix} a & b \\ c & d \end{pmatrix}\in G_{\mathbb{Z}}$ and any $z=(z_1,\ z_2)\in H\times H$,

$$\gamma(\mu(z)) = (cz_1+\ d)(c'z_2+\ d')\mu(\gamma(z)).$$

10.2. Now let us recall the definition of Weil representation and basic facts on theta series attached to the quadratic form Q.

Let $\mathcal{S}(L_{\mathbb{R}})$ be the space of Schwartz-Bruhat functions on $L_{\mathbb{R}}$. Then, following Weil [53], we define an action w of $SL_2(\mathbb{R})$ on $\mathcal{S}(L_{\mathbb{R}})$ by means of formulae:

$$w(\begin{pmatrix} a & 0 \\ 0 & a^{-1} \end{pmatrix})\Phi(v) = a\Phi(av), \quad (a\in\mathbb{R}-\{0\}),$$

$$w(\begin{pmatrix} 1 & b \\ 0 & 1 \end{pmatrix})\Phi(v) = \exp\{\pi ibQ(v)\}\Phi(v), \quad (b\in\mathbb{R}),$$

$$w(\begin{pmatrix} 0 & 1 \\ -1 & 0 \end{pmatrix})\Phi(v) = (-1)D^{3/2}\int_{L_{\mathbb{R}}} \exp\{2\pi iB(v,w)\}\Phi(v)dw,$$

for $\Phi(v)\in\mathcal{S}(L_{\mathbb{R}})$,
where dw is the normalized Lebesgue measure on $L_{\mathbb{R}}$ so that $\int_{L_{\mathbb{R}}/L} dw = 1$.

Though in general w defines only a projective representation of $SL_2(\mathbb{R})$, in our case w defines a true representation of $SL_2(\mathbb{R})$ by the result of Saito [39], because Q is of 4 (i.e. even) variables.

Put

$$\theta(\Phi) = \sum_{\ell \in L} \Phi(\ell), \quad \text{for any function } \Phi \text{ of } S(L_{\mathbb{R}}).$$

Then, the following invariance property of the distribution θ is known (cf. Proposition 1.1 and 1.6 of Shintani [49]).

10.3. Proposition. *For any* $\gamma = \begin{pmatrix} a & b \\ c & d \end{pmatrix} \in SL_2(\mathbb{Z})$ *with* $c \equiv 0 \bmod D$, *we have*

$$\theta(w(\gamma)\Phi) = \varepsilon_D(a)\theta(\Phi), \quad \textit{for any } \Phi \textit{ of } S(L_{\mathbb{R}}).$$

Here $\varepsilon_D(a)$ *is given by the Jacobi symbol* $\varepsilon_D(a) = \left(\frac{a}{D}\right)$.

To obtain the usual theta-formula, it is necessary to form functions of $S(L_{\mathbb{R}})$ with a good intertwining property with respect to w.

Fix a minimal majorant R of Q. For any point $\tau = s+it$ $(t>0)$ of H and $v \in L_{\mathbb{R}}$, we put

$$Q_\tau(v) = sQ(v) + itR(v).$$

Let $L_{\mathbb{R}} = L_{\mathbb{R}}^{+} \oplus L_{\mathbb{R}}^{-}$ be the unique direct sum decomposition of $L_{\mathbb{R}}$, such that

$$Q|_{L_{\mathbb{R}}^{+}} = R|_{L_{\mathbb{R}}^{+}}, \quad \text{and} \quad (-Q)|_{L_{\mathbb{R}}^{-}} = R|_{L_{\mathbb{R}}^{-}}.$$

Then there exists a vector w_c unique up to constant multiple such that

$$w_c \in X^{+} \cap (L_{\mathbb{R}}^{+} \otimes \mathbb{C}).$$

Let k be a non-negative integer. Consider functions $\Phi_k(v;\tau)$ and $\tilde{\Phi}_k(v;\tau)$ defined by

$$\Phi_k(v;\tau) = \{B(w_c, v)\}^k \exp\{\pi i Q_\tau(v)\},$$

and

$$\tilde{\Phi}_k(v;\tau) = t\Phi_k(v;\tau).$$

The following intertwining properties of Φ_k and $\tilde{\Phi}_k$ are well known (cf. Lemma 1.2, p.92 of [49]).

10.4. Proposition. *For any* $g = \begin{pmatrix} a & b \\ c & d \end{pmatrix} \in SL_2(\mathbb{R})$, *we have*

$$w(g)\Phi_k(v;\tau) = |c\tau+d|^2 (c\tau+d)^k \Phi_k\left(v; \frac{a\tau+b}{c\tau+d}\right),$$

and

$$w(g)\tilde{\Phi}_k(v;\tau) = (c\tau+d)^k \tilde{\Phi}_k\left(v; \frac{a\tau+b}{c\tau+d}\right).$$

§11. Construction of real Nebentype elliptic modular forms.

In this section, we construct real Nebentype elliptic modular forms whose Fourier coefficients are periods of Hilbert modualr forms, recalling some results of the previous paper [31], especially Theorem 1 of Section 3.

11.1. Let $f(z)$ be an element of $S_k(SL_2(O_F))$. Then, for $f(z)$ we can define a function $\Phi_f(g)$ on $G_{\mathbb{R}}$ by

$$\Phi_f(g)=j(g,\imath)^{-k}f(g(\imath)),$$

where

$$g=(g_1,\ g_2)=(\begin{pmatrix}a_1 & b_1\\ c_1 & d_1\end{pmatrix},\ \begin{pmatrix}a_2 & b_2\\ c_2 & d_2\end{pmatrix})\in G_{\mathbb{R}}=SL_2(\mathbb{R})\times SL_2(\mathbb{R}),$$

$\imath=(i,i)$ $H\times H$, and $j(g,\imath)=(c_1 i+d_1)(c_2 i+d_2)$.

Now for a fixed naural number k, we define a function Θ_Q on $H\times G_{\mathbb{R}}$ by

$$\Theta_Q(\tau;g)=\sum_{\ell\in L}\tilde{\Phi}_k(g^{-1}(\ell);\tau)$$

for the function $\tilde{\Phi}_k$ defined in Section 10.3.
Then by Proposition 10.3 and 10.4, we have

$$\Theta_Q(\frac{a\tau+b}{c\tau+d};g)=\varepsilon_D(a)(c\tau+d)^k\Theta_Q(\tau;g),$$

for any $\begin{pmatrix}a & b\\ c & d\end{pmatrix}\in\Gamma_0(D)$.

Form an integral

$$Sh_Q(f)(\tau)\underset{\text{dfn}}{=}\frac{(-i)(2\pi i)^{2(k-1)}}{(2\pi)^2}\int_{G_{\mathbb{Z}}\backslash G_{\mathbb{R}}}\Theta_Q(\tau;g)\Phi_f(g)dg.$$

Here dg is a Haar measure of $G_{\mathbb{R}}$.
This integral converges absolutely, and we can change the order of the summation of the series $\Theta_Q(\tau;g)$ and the integration, because $\Phi_f(g)$ is rapidly decreasing (since f is a cusp form), and because a majorant of the series $\Theta_Q(\tau;g)$ is slowly increasing in g by the estimate of p.117-118 of [31].

For the convenience of our computation in the late paragraphs, we fix a minimal majorant R of Q by

$$R(v)=D(a^2+\lambda^2+\lambda'^2+b^2)\quad\text{for } v=\begin{pmatrix}a & \lambda'\\ \lambda & b\end{pmatrix}\in L_{\mathbb{R}},$$

and put

$$w_c=\frac{1}{D}\begin{pmatrix}-1 & i\\ i & 1\end{pmatrix}.$$

Then, for $v=\begin{pmatrix}a & \lambda'\\ \lambda & b\end{pmatrix}\in L$,

$$B(w_c,v)=(a-b)+i(\lambda+\lambda'),$$

and moreover,

$$B(w_c,g^{-1}(v))=B(g(w_c),v)=j(g,\imath)B(\mu(g(\imath)),v)$$

for any $g\in G_{\mathbb{R}}$.

11.2. For any element v of $L_{\mathbb{R}}$, we denote by G_v the isotropy subgroup of $G_{\mathbb{R}}$ at $v\in L_{\mathbb{R}}$. For any $v\in L_{\mathbb{R}}$ such that $Q(v)\neq 0$, G_v is isomorphic to $SL_2(\mathbb{R})$.

Assume that $Q(v)=2m>0$ for $v\in L$. Choose a point z_0 of $H\times H$, and write $X_v(z_0)$ for its G_v-orbit in $H\times H$. Put $\Gamma_v=G_v\cap\Gamma$, and form an integral

$$W_v(f)\underset{dfn}{=}\int_{\Gamma_v\backslash X_v(z_0)}(2\pi i)^{2(k-1)}f(z)L(z,v)^{k-2}dz_1\wedge dz_2,$$

where $L(z,v)=B(\mu(z),v)=a+\lambda z_1+\lambda' z_2+bz_1z_2$ for $z=(z_1,\ z_2)\in H\times H$ and $v=\begin{pmatrix}a & \lambda'\\ \lambda & b\end{pmatrix}\in L$.

Then this integral converges and is independent of the choice of z_0 as shown in the proof (=Section 4) of [31].

11.3. <u>Theorem</u>. (<u>Theorem 1 of</u> [31], <u>Section</u> 3). <u>The integral $Sh_Q(f)$ converges absolutely and uniformly on any compact subset of H, and defines an element of $S_k(\Gamma_0(D),\ \varepsilon_D)$ for any $f\in S_k(SL_2(O_F))$. Moreover its Fourier expansion at infinity is given by</u>

$$Sh_Q(f)(\tau)=r\sum_{\{v\}}W_v(f)\exp[\pi iQ(v)],$$

<u>where the summation is taken over a complete representative system of the $G_{\mathbb{Z}}$-equivalence classes of elements v of L with $Q(v)>0$. Here r is a rational number independent of f.</u>

<u>Proof</u>. It suffices to check only the last statement that r is a rational number, because other statemants are already shown in [31], Theorem 1.

First, we normalize Haar measures in the following manner. Let K be the isotropy subgroup of G_R at $\imath=(i,i)\in H\times H$. Then $K\cong SO(2)\times SO(2)$. We normalize the Haar measure dk on K by

$$\int_K dk=(2\pi)^2.$$

Identifying $G_{\mathbb{R}}/K$ with $H\times H$, we normalize the Haar measure dg of $G_{\mathbb{R}}$ by

$$dg=\frac{dx_1dy_1}{y_1{}^2}\cdot\frac{dx_2dy_2}{y_2{}^2}\cdot dk,$$

where $g(i)=z=(z_1, z_2)$ with $z_i=x_i+\sqrt{-1}\,y_i$ $(i=1,2)$.
Let $K_{v,0}$ be the intersection $G_v\cap K_0$, where K_0 is the isotropy subgroup of z_0. Then $K_{v,0}\cong SO(2)$. We normalize the Haar measure dk_v of $K_{v,0}$ by

$$\int_{K_{v,0}} dk_v=2\ .$$

And for any $v\in L_{\mathbb{R}}$ with $Q(v)>0$, we normalize the Haar measure dg_v of G_v by

$$dg_v=dx_v dk_v,$$

where dx_v is a measure on $X_v(z_0)$ defined by the restriction of the 2-form on $H\times H$

$$\frac{i}{D}\,\frac{Q(v)}{L(z,v)^2}\,dz_1\wedge dz_2$$

to $X_v(z_0)$.

Under these normalization conditions, $Sh_Q(f)(\tau)$ is equal to

$$\frac{(-i)(2\pi i)^{2(k-1)}}{(2\pi)^2}\sum_{\{v\}}\int_{G_v\backslash G_{\mathbb{R}}} d\dot{g}\ \Big[\ \tilde{\Phi}_k(\tau;g^{-1}(v))\int_{\Gamma_v\backslash G_v}\Phi_f(g_vg)dg_v\Big].$$

By the argument of Section 4 of [31],

$$\int_{\Gamma_v\backslash G_v}\Phi_f(g_vg)dg_v=2\pi\int_{\Gamma_v\backslash G_v/K_{v,0}}\Phi_f(g_vg_0)dx_v$$

$$=2\pi L(i,g^{-1}(v))^{-k}\,\frac{iQ(v)}{D}\int_{\Gamma_v\backslash X_v/K_{v,0}} f(z)L(z,v)^{k-2}dz_1\wedge dz_2.$$

Note here that

$$j(g_vg,i)^{-k}=\frac{L(g_vg(i),v)^k}{L(i,g^{-1}(v))^k}$$

for any $g_v\in G_v$ and $g\in G_{\mathbb{R}}$.

Now let us calculate the integral

$$I=\int_{G_v\backslash G_{\mathbb{R}}} [\tilde{\Phi}_k(\tau;g^{-1}(v))\cdot 2\pi \frac{iQ(v)}{D} L(i,g^{-1}(v))^{-k}]d\dot{g}.$$

Since

$$\tilde{\Phi}_k(\tau;g^{-1}(v))= t\{B(w_c,g^{-1}(v))\}^k\exp[\pi i Q_\tau(g^{-1}(v))],$$

we have

$$t\Phi_k(\tau;g^{-1}(v))L(i,g^{-1}(v))^{-k}= t\exp[\pi i Q_\tau(g^{-1}(v))].$$

Therefore,

$$I=\frac{2\pi iQ(v)}{D}\int_{G_v\backslash G_{\mathbb{R}}} t\cdot\exp[\pi i Q_\tau(g^{-1}(v))]d\dot{g}.$$

The measure $d\dot{g}$ on $G_v\backslash G_{\mathbb{R}}$ is normalized by

$$dg= d\dot{g}dg_v \qquad (\dot{g}= G_v g).$$

We can identify the quotient

$$G_v\backslash G_{\mathbb{R}}\cong SO_0(2,1)\backslash SO_0(2,2)$$

with the space

$$S_{2,2}= \{(w_1,w_2,w_3,w_4)\in \mathbb{R}^4 \mid w_1^2+w_2^2-w_3^2-w_4^2=1\},$$

as in p.112 of [31].

Since $S_{2,2}$ is a homogenous space under $G_{\mathbb{R}}$, we can take $p_0=(1,0,0,0)$ as the fixed point of G_v.

Let $\mathfrak{g}$, $\mathfrak{g}_v$ be the Lie algebras of $G_{\mathbb{R}}$, and G_v, respectively. And let T_{p_0} be the tangent space of $S_{2,2}$ at p_0. Then via the Killing form of $\mathfrak{g}$, we have a decomposition

$$\mathfrak{g}=\mathfrak{g}_v \oplus \mathfrak{g}_v^{\perp},$$

and the orthogonal complement $\mathfrak{g}_v^{\perp}$ is naturally identified with T_{p_0}. By means of this identification, we have an equality of measures

$$d\dot{g}= 4\,\frac{dw_1\,dw_3\,dw_4}{|w_2|}=4\,\frac{dw_2\,dw_3\,dw_4}{|w_1|}.$$

Thus by a similar calculation as that of p.112 of [31] (cf. Formula (4.43)), we have

$$I=\frac{2\pi iQ(v)}{D}\cdot t\cdot\exp[\pi iQ(v)\tau]\cdot \mathrm{vol}(1)\cdot 2^{-2}\{tQ(v)\}^{-1}\cdot 4.$$

Since vol(1) is the volume of the unit sphere of dimension 1,

$$\mathrm{vol}(1)= 2\pi.$$

Therefore,

$$I=\frac{i(2\pi)^2}{D}\exp[\pi iQ(v)\tau].$$

Hence

$$Sh_Q(f)(\tau)=\frac{(2\pi i)^{2(k-1)}}{D}\sum_{\{v\}}\left\{\int_{\Gamma_v\backslash X_v(z_0)} f(z)L(z,v)^{k-2}dz_1\wedge dz_2\right\}\exp[\pi iQ(v)\tau]$$

which shows our theorem. q.e.d.

11.4. Corollary. *Especially when* $k=2$, *for any* $f\in S_2(SL_2(O_F))$,

$$Sh_Q(f)=r\sum_{\{v\}}\left(\int_{\Gamma_v\backslash X_v(z_0)}\omega_f\right)\exp[\pi iQ(v)\tau]$$

defines a holomorphic function on H *which belongs to* $S_2(\Gamma_0(D),\ \varepsilon_D)$. *Here* r *is a rational number independent of* f.

Proof. For k=2, we have

$$W_v(f)=\int_{\Gamma_v\backslash X_v(z_0)}(2\pi i)^2f(z)dz_1\wedge dz_2=\int_{\Gamma_v\backslash X_v(z_0)}\omega_f.$$

Hence, our corollary follows immediately from Theorem 11.3. q.e.d.

§12. Construction of Hilbert modular forms

Let $\Theta_Q(\tau;g)$ be the function on $H\times G_{\mathbb{R}}$ introduced in the previous section 11, and let $h(\tau)$ be an element of $S_k(\Gamma_0(D),\ \varepsilon_D)$. Then we form an integral

$$\int_{\Gamma_0(D)\backslash H} \overline{\Theta_Q(\tau;g)}h(\tau)t^k\,\frac{dsdt}{t^2} \qquad (\tau=s+it\in H)$$

and consider a function $DN_Q(h)(z)$ in $z\in H\times H=G_{\mathbb{R}}/K$, defined by

$$DN_Q(h)(g(i))=j(g,i)^k\int_{\Gamma_0(D)\backslash H} \overline{\Theta_Q(\tau;g)}h(\tau)t^k\,\frac{dsdt}{t^2}.$$

We want to show that this function is an element of $S_k(SL_2(O_F))$ in this section.

In order to prove that $DN_Q(h)$ is a holomorphic Hilbert modular form of weight k, and to show that this construction of Hilbert modular forms from real Nebentype elliptic modular forms coincides with that of Naganuma [29], we have to investigate the relation between the Fourier coefficients of h and $DN_Q(h)$. It is possible to modify the method of Niwa [30] to see this relation, but to make our exposition short, we choose a method to reduce our problem to the results of Zagier [55].

Remark. Similar type of liftings by using theta series attached to indefinite quadratic forms are also discussed by Asai [3] and Rallis-Schiffmann [35] in detail. See also the references of [31] on the related works of Kudla, Rallis-Schiffmann, and Vigneras.

12.1. Now our task is to check that our lifting $DN_Q(h)$ coincides with that of Zagier [55]. Define $\Omega(\tau;z)$ by

$$\Omega(\tau;z)=\sum_{v\in L,\ Q(v)>0} (\tfrac{1}{2}Q(v))^{k-1}L(z,v)^{-k}\exp[\pi iQ(v)\tau],$$

as in [55]. We want to show an identity

$$DN_Q(h)(z)=\frac{r}{\pi}\int_{\Gamma_0(D)\backslash H}\Omega(-\bar{\tau};z)h(\tau)t^k\,\frac{dsdt}{t^2}$$

for any element h of $S_2(\Gamma_0(D),\ \varepsilon_D)$, where r is a rational constant independent of h.

As noted in a remark of p.119 of [31], the function

$$\frac{k}{4\pi}\,\Omega(-\overline{\tau}_2;z)$$

is the lifting via DN_Q of the reproducing kernel $K_k^{D,\varepsilon}(\tau,\tau_2)$ of $S_k(\Gamma_0(D),\ \varepsilon_D)$ as a function in τ. Therefore the above identity is almost evident, although we have to pay special attention to the case of weight k=2, because of a trouble of the convergence of the series in question. But the reason why such a trouble of convergence occurs is not related to the nature of $\Theta_Q(\tau;g)$, but due to the fact that the reproducing kernel $K_k^{D,\varepsilon}$ of $S_k(\Gamma_0(D),\ \varepsilon_D)$ does not converges absolutely for k=2, because the corresponding discrete series representation of $SL_2(\mathbb{R})$ is not integrable.

In fact, as we see in the next paragraph, the integral

$$\int_{\Gamma_0(D)\backslash H} \overline{\Theta_Q(\tau;g)}h(\tau)t^k\,\frac{dsdt}{t^2}$$

itself converges absolutely, because $|h(\tau)|$ is rapidly decreasing and a majorant of $\Theta_Q(\tau;g)$ is slowly increasing.

12.2. Lemma. The integral

$$\int_{\Gamma_0(D)\backslash H} \overline{\Theta_Q(\tau;g)}h(\tau)t^k\,\frac{dsdt}{t^2}$$

converges absolutely for any cusp form $h(\tau)\in S_k(\Gamma_0(D),\ \varepsilon_D)$.

Proof. For any cusp $r=a/b$ ($(a,b)=1$, $b>0$) or $r=\infty$ of $\mathbb{Q}\cup\{\infty\}$, and any point $\tau=s+it$ ($t>0$) of H, we define a "distance" $d(\tau,r)$ by

$$d(\tau,r)=\frac{t}{(bs-a)^2+b^2t^2}\qquad \text{for } r=a/b\neq\infty,$$

and

$$d(\tau,\infty)=t.$$

Then $d(\tau,r)$ has an invariance property:

$$d(\gamma(\tau),\gamma(r))=d(\tau,r)\qquad \text{for any } \gamma\in SL_2(\mathbb{Z}).$$

Put

$$H_M=\{\tau\in H\ |\ d(\tau,r)\leqq M\},$$

and

$$\Sigma_r=\{\tau\in H\ |\ d(\tau,r)\geqq M\}.$$

Then for a sufficiently large real number M,

$$\Sigma_r\cap\gamma\Sigma_{r'}=\phi\qquad \text{for any } \gamma\in\Gamma_0(D),$$

if r and r' are not equivalent with respect to $\Gamma_0(D)$, and for each cusp r,

$$\Sigma_r \cap \gamma\Sigma_r \neq \phi, \text{ for } \gamma \in \Gamma_0(D),$$

if and only if $\gamma(r)=r$.

Therefore for a sufficiently large real number M, we have

$$\Gamma_0(D)\backslash H = \Gamma_0(D)\backslash H_M \cup \left(\bigcup_{i=1}^{s} \Gamma_i \backslash \Sigma_{r_i}\right),$$

where $\{r_1, \ldots, r_s\}$ is a complete representative system of the $\Gamma_0(D)$-equivalence classes of cusps, and $\Gamma_i = \{\gamma \in \Gamma_0(D) \mid \gamma(r_i)=r_i\}$.

Since $\Gamma_0(D)\backslash H_M$ is a compact set, it is sufficient to check the convergence of the integral on each subset $\Gamma_i\backslash\Sigma_{r_i}$ of $\Gamma_0(D)\backslash H$. Recall that all cusps of $\Gamma_0(D)$ are equivalent under $SL_2(\mathbf{Z})$, and that the transformation of $\Theta_Q(\tau;g)$ with respect to any element $\gamma = \begin{pmatrix} a & b \\ c & d \end{pmatrix} \in SL_2(\mathbf{Z})$ is given by

$$\Theta_Q\left(\frac{a\tau+b}{c\tau+d};g\right) = (c\tau+d)^k \sum_{h\in L^*/L} c_\gamma(h)\Theta_Q(\tau;g;h).$$

Here $c_\gamma(h)$ are constants depending only on γ, and here $\Theta_Q(\tau;g;h)$ is defined by

$$\sum_{\ell\in L^*,\ \ell\equiv h \bmod L} \tilde{\Phi}_k(g^{-1}(\ell);\tau),$$

for $h\in L^*$ (cf. Proposition 1.6 of Shintani [49]). We can estimate these $\Theta_Q(\tau;g;h)$ similarly as $\Theta_Q(\tau;g)=\Theta_Q(\tau;g;0)$.

Hence to show the convergence of the integral, it suffices to show it for

$$\int_{\Gamma_\infty\backslash\Sigma_\infty} \overline{\Theta_Q(\tau;g)}h(\tau)t^k \frac{dsdt}{t}.$$

By an estimate of Section 5.2 of [31],

$$\sum_{v\in L} |\tilde{\Phi}_k(g^{-1}(v);\tau)| \leqq t \sum_{v\in L} \|v\|^k \cdot \|g\|^{M''} \cdot \exp\left(-c' \frac{\|v\|^2 t}{\|g\|^{M'}}\right)$$

for some positive real numbers c', M', and M" (cf. Formula (5.30) of [31]). On the other hand,

$$h(\tau) = O(t^{-k/2}).$$

Since

$$\sum_{v\in L} \|v\|^k \exp(-c'\frac{\|v\|^2 t}{\|g\|^{M'}}) \sim \text{const.}\int_0^\infty r^{k+4-1}\exp(-c'\frac{r^2 t}{\|g\|^{M'}})dr$$

$$\sim \text{const. } t^{-(k+4)}\|g\|^{M'(k+4)},$$

the integral in question converges for a sufficiently large M. q.e.d.

12.3. Let us recall the reproducing kernel of $S_k(\Gamma_0(D), \varepsilon_D)$. If k is an even integer with $k\geqq 4$, the reproducing kernel is given by

$$K_k^{D,\varepsilon}(\tau_1,\tau_2)=\frac{(k-1)(2i)^k}{4\pi}\ \frac{1}{2}\sum_{\gamma=\binom{a\ b}{c\ d}\in\Gamma_0(D)} \varepsilon_D(a)\{\tau_1-\gamma(\bar{\tau}_2)\}^{-k}(c\bar{\tau}_2+d)^{-k}$$

$$=\sum_{m=1}\overline{G_m(\tau_2)}\exp[2\pi im\tau_1]=\sum_{m=1}G_m(\tau_1)\overline{\exp[2\pi im\tau_2]}$$

with

$$G_m(\tau)=\frac{(4\pi m)^{k-1}}{(k-2)!}\ \frac{1}{2}\sum_{\gamma=\binom{a\ b}{c\ d}\in\Gamma_\infty\backslash\Gamma_0(D)}\varepsilon_D(a)(c\tau+d)^{-k}\exp[2\pi im\gamma(\tau)],$$

where $\Gamma_\infty=\{\pm\binom{1\ b}{0\ 1}\mid b\in\mathbb{Z}\}$.
In this case, for $z=g(i)$ of $H\times H$

$$DN_Q(K_k^{D,\varepsilon}(\tau,\tau_2))(z)=j(g,i)^k\int_{\Gamma_0(D)\backslash H}\overline{\Theta_Q(\tau;g)}K_k^{D,\varepsilon}(\tau,\tau_2)t^k\frac{dsdt}{t^2}$$

is given by

$$\sum_{m=1}DN_Q(G_m)(z)\overline{\exp[2\pi im\tau_2]}.$$

Similarly as in Section 5.1 (p.115) of [31],

$$DN_Q(G_m)(z)=\frac{(4\pi m)^{k-1}}{(k-2)!}j(g,i)^k\int_0^\infty\int_0^1\overline{\Theta_Q(\tau;g)}\exp[2\pi im\tau]t^{k-2}dsdt$$

$$=\frac{(4\pi m)^{k-1}}{(k-2)!}j(g,i)^k\sum_{\substack{v\in L\\ Q(v)=2m}}\int_0^\infty\overline{B(w_c,g^{-1}(v))}^k\exp[-\pi t(Q+R)(g^{-1}(v))]t^{k-1}dt$$

$$=\frac{(4\pi m)^{k-1}}{(k-2)!}\ \frac{(k-1)!}{\pi^k}\ j(g,i)^k\sum_{\substack{v\in L\\ Q(v)=2m}}\frac{\overline{B(w_c,g^{-1}(v))}^k}{\{(Q+R)(g^{-1}(v)\}^k}$$

$$= \frac{(k-1)}{\pi}(4m)^{k-1} j(g,i) \sum_{\substack{v\in L\\ Q(v)=2m}} \left[\frac{|B(w_c,g^{-1}(v))|^{2k}}{\{(Q+R)(g^{-1}(v))\}^k} \times B(w_c,g^{-1}(v))^{-k}\right].$$

For any vector $v=\begin{pmatrix} a & \lambda' \\ \lambda & b\end{pmatrix}$ of $L_{\mathbb{R}}$, we have

$$\frac{|B(w_c,v)|^{2k}}{\{(Q+R)(v)\}^k} = \frac{|(a-b)+i(\lambda+\lambda')|^{2k}}{\{-2Dab+2D\lambda\lambda'+D(a^2+\lambda^2+\lambda'^2+b^2)\}^k} = D^{-k}$$

Hence,

$$DN_Q(G_m)(z) = \frac{(k-1)}{\pi}(4m)^{k-1}D^{-k} \sum_{\substack{v\in L\\ Q(v)=2m}} B(\mu(g(i)),v)^{-k}$$

$$= \frac{(k-1)}{\pi}(4m)^{k-1}D^{-k} \sum_{\substack{v\in L\\ Q(v)=2m}} L(z,v)^{-k},$$

with $L(z,v)= a+\lambda z_1+\lambda' z_2+bz_1z_2$.

Thus

$$DN_Q(K_k^{D,\varepsilon}(\ ,\tau))(z) = \frac{(k-1)}{\pi}\frac{4^{k-1}}{D^k}\sum_{m=1} m^{k-1}\{\sum_{\substack{v\in L\\ Q(v)=2m}} L(z,v)^{-k}\} \exp[2\pi i m\tau]$$

$$= \frac{(k-1)}{\pi}\frac{4^{k-1}}{D^k}\,\Omega(-\bar{\tau};z),$$

where

$$\Omega(\tau;z) = \sum_{m=1} m^{k-1}\Big(\sum_{\substack{v\in L\\ Q(v)=2m}} L(z,v)^{-k}\Big)\exp[2\pi i m\tau]$$

is the kernel function of Zagier [55].

Since $K_k^{D,\varepsilon}(\tau_1,\tau_2)$ is the reproducing kernel of $S_2(\Gamma_0(D),\varepsilon_D)$, we have

$$DN_Q(h)(z) = \frac{(k-1)}{\pi}\frac{4^{k-1}}{D^k}\int_{\Gamma_0(D)\backslash H} \Omega(-\bar{\tau};z)h(\tau)t^{k-2}dsdt.$$

When k=2, we consider

$$G_m(\tau) = \lim_{\sigma\to 0} G_{m,\sigma}(\tau)$$

with

$$G_{m,\sigma}=\frac{(4\pi m)^{k-1}}{(k-2)!}\frac{1}{2}\sum_{\gamma=\left(\begin{smallmatrix}a&b\\c&d\end{smallmatrix}\right)\ \Gamma_\infty\backslash\Gamma_0(D)}\varepsilon_D(a)(c\tau+d)^{-2}|c\tau+d|^{-2\sigma}\exp[2\pi i m\gamma(\tau)],$$

and form an integral

$$\int_{\Gamma_0(D)\backslash H}\overline{\Theta_Q(\tau;g)}G_{m,\sigma}(\tau)t^{2+\sigma}\frac{dsdt}{t^2}\qquad\text{for } \sigma>0,$$

which is equal to

$$\int_{\Gamma_\infty\backslash H}\overline{\Theta_Q(\tau;g)}\exp[2\pi i m\tau]t^{2+\sigma}\frac{dsdt}{t^2}$$

$$=\frac{(k-1)}{4\pi}(4m)^{k-1}D^{-(k+\sigma)}j(g,i)^{-k}\sum_{\substack{v\in L\\Q(v)=2m}}L(z,v)^{-k}\times|(Q+R)(g^{-1}(v))|^{-2\sigma}$$

$$=\frac{(k-1)}{4\pi}(4m)^{k-1}D^{-k-\sigma}j(g,i)^{-k}\sum_{\substack{v\in L\\Q(v)=2m}}L(z,v)^{-k}|L(z,v)|^{-2\sigma}\times|j(g,i)|^{-2\sigma}.$$

Applying $\lim_{\sigma\to 0}$ to the both hand sides of the above formula, we can show that $DN_Q(G_m)(z)$ is a Hilbert modular cusp form of weight 2 (cf. Appendix, Theorem 2 of [55]).

Since the m-th Fourier coefficient of $\Omega(\tau;z)$ as a function in τ is given by

$$\frac{(k-2)!}{(4\pi m)^{k-1}}DN_Q(G_m)(z),$$

we have

$$DN_Q(G_m)(z)=\frac{(k-1)}{\pi}\frac{4^{k-1}}{D^k}\int_{\Gamma_0(D)\backslash H}\Omega(-\bar{\tau};z)G_m(\tau)t^{k-2}dsdt$$

for any G_m. Hence similar identity is valid for any h of $S_2(\Gamma_0(D),\varepsilon_D)$ in place of G_m in the above identity, because $\{G_m \mid m=1,2,\dots\}$ spans $S_2(\Gamma_0(D),\varepsilon_D)$.

Comparing with the results of Zagier [55] (especially Theorem 5, and some formulae before this theorem), we have the following theorem.

12.4. Theorem. Assume that D is a prime with $D\equiv 1 \bmod 4$, and that the class number of $F=\mathbb{Q}(\sqrt{D})$ is 1. Then we have the following:

(i) Let

$$h(\tau)=\sum_{n=1} a_n \exp[2\pi i n\tau] \qquad (a_1=1)$$

be a normalized primitive form of $S_k(\Gamma_0(D),\ \varepsilon_D)$. Then,

$$f(z)=2^{-(k+1)}D^k(-1)^{k/2}\,DN_Q(h)(z)$$

is a normalized primitive form of $S_k(SL_2(O_F))$. Moreover, let

$$f(z)=\sum_{\nu\in O_{F,+}} a((\nu))\exp[2\pi i(\nu\omega z_1+\nu'\omega' z_2)]$$

be the Fourier expansion of f at infinity, then

$$L^{(1)}(s,f/F)=L(s,h)L(s,h^\rho),$$

where

$$L(s,h)=\sum_{n=1}^{\infty} a_n n^{-s},\quad L(s,h^\rho)=\sum_{n=1}^{\infty}\rho(a_n)n^{-s}=\sum_{n=1}^{\infty}\bar{a}_n n^{-s},$$

and

$$L^{(1)}(s,f/F)=\sum_{\substack{\mathfrak{n};\text{integral}\\ \text{ideals of } O_F}} a(\mathfrak{n})N_{F/\mathbb{Q}}(\mathfrak{n})^{-s}.$$

(ii) Define a subspace $S_k^+(\Gamma_0(D),\ \varepsilon_D)$ of $S_k(\Gamma_0(D),\ \varepsilon_D)$ by

$$S_k^+(\Gamma_0(D),\ \varepsilon_D)=\{\ h=\sum_{n=1} a_n\exp[2\pi i n\tau]\in S_k(\Gamma_0(D),\ \varepsilon_D)\ |\ a_n=0,\ \text{if } (\tfrac{n}{D})=-1\}.$$

Then the restriction of DN_Q to $S_k^+(\Gamma_0(D),\ \varepsilon_D)$ is injective. And if we put

$$S_k^-(\Gamma_0(D),\ \varepsilon_D)=\{\ h=\sum_{n=1} a_n\exp[2\pi i n\tau]\in S_k(\Gamma_0(D),\ \varepsilon_D)\ |\ a_n=0,\ \text{if } (\tfrac{n}{D})=+1\ \},$$

$DN_Q(h)=0$, for any $h\in S_k^-(\Gamma_0(D),\ \varepsilon_D)$.

Remark. Though the injectivity of DN_Q on $S_k^+(\Gamma_0(D),\ \varepsilon_D)$ is claimed in Proposition 1 of Section 5 of [55] without proof, the proof is not difficult (at least for prime D) as remarked there. This injectivity also follows from Theorem 3 (p.168) of Saito [38].

12.5. Corollary of Theorem 12.4. Let h be a normalized primitive form

<u>of</u> $S_k(\Gamma_0(D), \varepsilon_D)$, <u>and put</u>

$$f = 2^{-(k+1)} D^k (-1)^{k/2} DN_Q(h).$$

<u>Let</u>

$$\chi : \mathbf{Z} \longrightarrow \mathbf{C}^\times$$

<u>be a Dirichlet character modulo</u> m. <u>Then we have an identity</u>

$$L^{(1)}(s, f/F, \chi\circ N_{F/\mathbf{Q}}) = L(s,h,\chi)L(s,h^\rho,\chi) \quad \text{for } \mathrm{Re}(s) > k,$$

<u>where</u>

$$L(s,h,\chi) = \sum_{n=1} a_n \chi(n) n^{-s}, \quad L(s,h^\rho,\chi) = \sum_{n=1} \rho(a_n)\chi(n) n^{-s},$$

<u>and</u>

$$L^{(1)}(s, f/F, \chi\circ N_{F/\mathbf{Q}}) = \sum_{\mathfrak{n};\, \text{ideals of } O_F} a(\mathfrak{n})\chi(N_{F/\mathbf{Q}}(\mathfrak{n}))N_{F/\mathbf{Q}}(\mathfrak{n})^{-s}.$$

<u>Proof</u>. This corollary follows immediately from the part (i) of Theorem 12.4. q.e.d.

12.6. <u>Remark</u>. By the identity

$$L^{(1)}(s, f/F) = L(s,h)L(s,h^\rho),$$

we can see that the field of eigenvalues K_f of the lifted form $f = DN_Q(h)$ is the maximal totally real subfield

$$k_h = \{\ a \in K_h \mid \rho_h(a) = a\ \}$$

of the field of eigenvalues K_h of h, i.e.

$$K_f = k_h.$$

§13. The adjointness formula.

In this section, we discuss the adjointness of the mappings

$$DN_Q: S_k(\Gamma_0(D), \varepsilon_D) \longrightarrow S_k(SL_2(O_F))$$

and the mapping

$$Sh_Q: S_k(SL_2(O_F)) \longrightarrow S_k(\Gamma_0(D), \varepsilon_D)$$

with respect to the Petersson metrics.

13.1. Form an integral

$$I(f;h)=\int_{\Gamma_0(D)\backslash H}\{\int_{G_{\mathbb{Z}}\backslash G_{\mathbb{R}}}\Theta_Q(\tau;g)\Phi_f(g)\overline{h(\tau)}dg\}t^k\,\frac{dsdt}{t^2}$$

for $f\in S_k(SL_2(O_F))$ and $h\in S_k(\Gamma_0(D), \varepsilon_D)$.

As we have seen already, a majorant of $\Theta_Q(\tau;g)$ given by

$$\sum_{\{v\}}|\tilde{\Phi}_k(g^{-1}(v);\tau)|$$

is a slowly increasing function in τ and g, and $\Phi_f(g)$ and $h(\tau)$ are rapidly decreasing, because $\Phi_f(g)$ and $h(\tau)$ are cusp forms. Therefore the above integral $I(f;h)$ converges absolutely. Hence by the theorem of Fubini, we have

$$I(f;h)=r\,\frac{i(2\pi)^2}{(2\pi i)^{2(k-1)}}\int_{\Gamma_0(D)\backslash H}Sh_Q(f)(\tau)\overline{h(\tau)}t^k\,\frac{dsdt}{t^2}$$

and

$$I(f;h)=r'(2\pi)^2\int_{SL_2(O_F)\backslash(H\times H)}f(z)\overline{DN_Q(h)(z)}(y_1y_2)^k\,\frac{dx_1dy_1}{y_1^2}\,\frac{dx_2dy_2}{y_2^2},$$

where r and r' are rational numbers independent of f and h.

13.2. Proposition. (The adjointness formula). *For any element h of $S_k(\Gamma_0(D), \varepsilon_D)$ and any element f of $S_k(SL(O_F))$, we have*

$$r(Sh_Q(f), h)=(f, DN_Q(h))$$

with respect to the Petersson metrics (,). Here r is a rational number independent of f and h, and here the Petersson metrics are normalized by

$$(h, h')=(-2i)\int_{\Gamma_0(D)\backslash H}(2\pi i)^{2(k-1)}h(\tau)\overline{h'(\tau)}t^k\,\frac{dsdt}{t^2}$$

for elliptic modular cusp forms h, h' $\in S_k(\Gamma_0(D), \varepsilon_D)$,
and by

$$(f, f') = (-2i)^2 \int_{SL_2(O_F)\backslash(H\times H)} (2\pi i)^{4(k-1)} f(z)\overline{f'(z)}(y_1 y_2)^k \frac{dx_1 dy_1 dx_2 dy_2}{y_1^2 y_2^2}$$

for Hilbert modular cusp forms f, f' $\in S_k(SL_2(O_F))$.

Especially when k=2, we have

13.3. Proposition. (The adjointness formula for k=2). For any h $S_2(\Gamma_0(D), \varepsilon_D)$ and any f $S_2(SL_2(O_F))$, we have

$$r\int_C \omega_{Sh_Q(f)} \wedge \overline{\omega}_h = \int_S \omega_f \wedge \overline{\omega_{DN_Q(h)}}.$$

Here r is a rational constant independent of f and h, and here C is the modular curve $(\Gamma_0^{\varepsilon}(D)\backslash H)^*$ with $\Gamma_0^{\varepsilon}(D)$ the kernel of the mutiplicator

$$\varepsilon_D : \begin{pmatrix} a & b \\ c & d \end{pmatrix} \in \Gamma_0(D) \longmapsto \left(\frac{a}{D}\right) \in \{\pm 1\}$$

of $\Gamma_0(D)$.
Proof. Since

$$\int_C \omega_{Sh_Q(f)} \wedge \overline{\omega}_h = 2\int_{\Gamma_0(D)\backslash H} \omega_{Sh_Q(f)} \wedge \overline{\omega}_h,$$

our proposition follows immediately from Proposition 13.2. q.e.d.

As a consequence of Theorem 12.4 and the adjointness formula (Prop. 13.2), we have the following proposition.

13.4. Proposition. If f is a primitive form of $S_k(SL_2(O_F))$, which does not belong to the image of the mapping DN_Q, then

$$Sh_Q(f)=0,$$

and a fortiori the sum of integrals

$$\sum_{\substack{\{v\} \\ Q(v)=2m}} \int_{\Gamma_v\backslash X_v(z_0)} f(z)L(z,v)^{k-2} dz_1 \wedge dz_2$$

is zero for each m (m=1,2,...).

Proof. The space $S_k(\Gamma_0(D), \varepsilon_D)$ is spanned by primitive forms. By Theorem 12.4, the image $DN_Q(h)$ of any primitive form h in $S_k(\Gamma_0(D), \varepsilon_D)$ is a primitive form of $S_k(SL(O_F))$. Therefore any element of the image of $S_k(\Gamma_0(D), \varepsilon_D)$ via the mapping DN_Q is a linear combination of primitive forms of the form $DN_Q(h)$ with some primitive forms h of $S_k(\Gamma_0(D), \varepsilon_D)$.

If a primitive form f of $S_k(SL_2(O_F))$ does not belong to the image $DN_Q(S_k(\Gamma_0(D), \varepsilon_Q))$, then for any primitive $h \in S_k(\Gamma_0(D), \varepsilon_D)$, f and $DN_Q(h)$ are two primitive forms with different eigenvalues for some Hecke operators. Therefore these two forms f and $DN_Q(h)$ are orthogonal with respect to the Petersson metric:

$$\int_{SL_2(O_F)\backslash(H\times H)} f(z)\overline{DN_Q(h)(z)}(y_1y_2)^{k-2}dx_1dy_1dx_2dy_2 = 0.$$

Hence by the adjointness formula (Proposition 13.2),

$$\int_{\Gamma_0(D)\backslash H} Sh_Q(f)(\tau)\overline{h(\tau)}t^{k-2}dsdt = 0$$

for any primitive form h of $S_k(\Gamma_0(D), \varepsilon_D)$. Therefore $Sh_Q(f)=0$, because the set of primitive forms spans $S_k(\Gamma_0(D), \varepsilon_D)$.

The latter part of the proposition follows immediately from the Fourier expansion of $Sh_Q(f)$ (cf. Theorem 11.3). q.e.d.

When f is obtained by the lifting DN_Q, we have the following.

13.5. <u>Proposition</u>. <u>Assume that the discriminant</u> D <u>of</u> F <u>is a prime number with</u> $D\equiv 1$ <u>mod</u> 4. <u>If</u> $f=DN_Q(h)$ <u>for some primitive form</u> h <u>of</u> $S_k(\Gamma_0(D), \varepsilon_D)$. <u>Then</u>

$$Sh_Q(f)=a(h)(h+h^\rho).$$

<u>Here</u> a(h) <u>is a non-zero constant depending only on</u> h <u>given by</u>

$$a(h)=r(2\pi i)^{2(k-1)}\sum_{\substack{\{v\}\\ Q(v)=2}} \int_{\Gamma_v\backslash X_v(z_0)} f(z)L(z,v)^{k-2}dz_1\wedge dz_2 \neq 0,$$

<u>where</u> r <u>is a rational number independent of</u> h.

<u>Proof</u>. By Theorem 12.4, DN_Q is an injective linear mapping from $S_k^+(\Gamma_0(D), \varepsilon_D)$ to $S_k(SL_2(O_F))$, and for any normalized primitive form h, we have $DN_Q(h)=DN_Q(h^\rho)$ for the automorphism ρ. Therefore $(h+h^\rho)/2$ is the unique form in $S_k^+(\Gamma_0(D), \varepsilon_D)$ such that $r_0DN_Q((h+h^\rho)/2)$ is the normalized primitive form $f=r_0DN_Q(h)$, where

$$r_0=(-1)^{k/2}\cdot 2^{-(k+1)}D^k.$$

Therefore if h' is another normalized primitive form of $S_k(\Gamma_0(D), \varepsilon_D)$ such that $h'\neq h$ and $h'\neq h^\rho$, then $f'=r_0DN_Q(h')$ is a primitive form of

$S_k(SL_2(O_F))$ with eigenvalues different from those of f for some Hecke operators. Hence

$$\int_S f(z)\overline{f'(z)}(y_1y_2)^{k-2}dx_1dy_1dx_2dy_2=0.$$

By the adjointness formula (13.2), this implies that

$$\int_{\Gamma_0(D)\backslash H} Sh_Q(f)(\tau)\overline{h'(\tau)}t^{k-2}dsdt=0.$$

Therefore, $Sh_Q(f)$ is a linear combination of h and h^ρ. Since the Fourier coefficients C_m of the Fourier expansion

$$Sh_Q(f)(\tau)=\sum_{m=1} C_m\exp[2\pi im\tau]$$

are zeros for m with $(\frac{m}{D})=-1$ by the definition of the quadratic form Q, the form $Sh_Q(f)$ belongs to $S_k^+(\Gamma_0(D), \varepsilon_D)$. Accordingly, $Sh_Q(f)$ is a constant multiple of $h+h^\rho$. The fact that a(h) is not zero follows immediately from the adjointness formula (Proposition 13.2):

$$a(h)(h, h)=(Sh_Q(f), h)=(f, DN_Q(h))=(f, f)\neq 0.$$

Assume that D is a prime number with $D\equiv 1$ mod 4, and let us prove the last part of our proposition. Since h and h are normalized, to determine a(h) it suffices to calculate the first Fourier coefficient of $Sh_Q(f)(\tau)$. Thus we can conclude the proof of our proposition by Theorem 11.3. q.e.d.

13.6. Remark. Especially when k=2, the constant a(h) in Proposition 13.5 is given by

$$a(h)=r\sum_{\substack{\{v\}\\ Q(v)=2}}\left\{\int_{\Gamma_v\backslash X_v(z_0)}\omega_f\right\}\neq 0.$$

13.7. Remark. As refered in the introduction of the previous paper [31], our result in this chapter overlaps substantially with that of Hirzebruch-Zagier [19],[56]. Let us explain this point more precisely.

Let N be a positive integer, and let v be any vector of L such that $Q(v)=-2N$. Let G_v be the isotropy subgroup of G_R at v. And for some point z_0 of $H\times H$ consider the G_v-orbit $Y_v(z_0)$ of z_0. Then it is easy to check that $Y_v(z_0)$ is a complex analytic submanifold of $H\times H$. Thus we can regard the quotient $\Gamma_v\backslash Y_v(z_0)$ as an algebraic curve on S.

Consider a sum of algebraic cycles

$$\sum_{\substack{\{v\} \\ Q(v)=-2N}} \Gamma_v \backslash Y_v(z\ )$$

as a cycle on Hilbert modular surface S. Then this cycle is homologous to the curve F_N of [19], [56]. Similarly as in Section 11, we ca show that

$$\sum_{N=1} \{\int_{F_N} \eta_{f,1}\}\exp[2\pi i N\tau], \text{ or } \sum_{N=1} \{\int_{F_N} \eta_{f,2}\}\exp[2\pi i N\tau]$$

defines an element of $S_2(\Gamma_0(D), \varepsilon_D)$ for any $f \in S_2(SL_2(O_F))$. Here, $\eta_{f,1}$ and $\eta_{f,2}$ are (1,1) type forms defined by

$$\eta_{f,1}=(2\pi i)^2 f(\varepsilon_0 z_1, \varepsilon_0' \overline{z_2})dz_1 \wedge \overline{dz_2},\ \eta_{f,2}=(2\pi i)^2 f(\varepsilon_0' \overline{z_1}, \varepsilon_0 z_2)\overline{dz_1} \wedge dz_2,$$

where ε_0 is the fundamental unit of O_F, satisfying $\varepsilon_0>0$ and $\varepsilon_0'<0$. Since F_N is an algebraic cycle,

$$\int_{F_N} \omega_f = \int_{F_N} \overline{\omega}_f = 0.$$

Therefore we have

$$\sum_{N=1} \{\int_{F_N} \omega\ \}\exp[2\pi i N\tau] \in S_2(\Gamma_0(D), \varepsilon_D)$$

for arbitrary $\omega \in H^2_{sp}(S,\mathbb{C})$. On the other hand, the Eisenstein-Siegel formula for the theta series of the quadratic form Q implies that

$$\sum_{N=1} \{\int_{F_N} (\eta_1+\eta_2)\}\exp[2\pi i N\tau]$$

is an Eisenstein series of weight 4/2=2 with respect to $\Gamma_0(D)$ with multiplicator ε_D, and

$$\int_{F_N} (\eta_1-\eta_2)=0 \quad \text{for any N.}$$

Therefore for any $\omega \in W_2H^2(S,\mathbb{C})$, the series

$$\sum_{N=1} \{\int_{F_N} \omega\ \}\exp[2\pi i N\tau]$$

is a modular form of weight 2 with respect to $\Gamma_0(D)$ with multiplicator ε_D. We may interpret these period integrals

$$\int_{F_N} \omega$$

of the Fourier coefficents as the intersection numbers of F_N and the Poincaré dual of ω. Thus we can show the main results of [19], [56].

Chapter IV. Period relation for the lifting of modular forms and transcendental cycles.

§14. Hodge structures attached to real Nebentype elliptic modular forms of weight 2.

In this section, we discuss the period relation of Riemann (Proposition 14.5) for real Nebentype elliptic modular cusp forms of weight 2. More precisely speaking, we formulate the period relation of Riemann for the Hodge structures attached real Nebentype primitive forms of weight 2. Because the category of polarized Hodge structures of weight 1 is equivalent to the opposite category of polarized abelian varieties, the formalism attaching Hodge structures of weight 1 to real Nebentype elliptic primitive cusp forms of weight 2 follows immediately from the theory of Shimura [43]. Therefore we omit some of details and proofs of this formalism. About basic facts and definition, we refer to [43].

14.1. Suppose that D is the discriminant of a real quadratic field F. From now on we assume that D is a prime number and that the class number of F is 1 for simplicity.

We consider a congruence subgroup

$$\Gamma_0(D)=\{\begin{pmatrix} a & b \\ c & d \end{pmatrix}\in SL_2(\mathbb{Z}) \mid c\equiv 0 \bmod D\}$$

of $SL_2(\mathbb{Z})$, which acts on the complex upper half plane H in the usual manner. Let ε_D be a character $\varepsilon_D:\mathbb{Z}\longrightarrow\{\pm 1\}$ defined by $\varepsilon_D(a)=(\frac{D}{a})$, where $(-)$ is the Jacobi symbol. We denote by $\Gamma_0^{\varepsilon}(D)$ the kernel of a character of $\Gamma_0(D)$ defined by

$$\begin{pmatrix} a & b \\ c & d \end{pmatrix}\in\Gamma_0(D)\longmapsto\varepsilon_D(a).$$

Evidently $\Gamma_0^{\varepsilon}(D)$ is a subgroup of $\Gamma_0(D)$ of index 2.

Let C be the standard compactification $(\Gamma_0^{\varepsilon}(D)\backslash H)^*$ of the quotient Riemann surface $\Gamma_0^{\varepsilon}(D)\backslash H$. Then C has a canonical model over $\mathbb{Q}$ (cf. Chapter 7 of [43]). An involutive automorphism of H defined by

$$z \longmapsto -\frac{1}{Dz} \quad (\text{for } z \in H)$$

belongs to the normalizer of $\Gamma_0^{\varepsilon}(D)$. Therefore, on passing to the quotient C, this automorphism induces an involutive automorphism w_D of C.

A non-holomorphic involutive automorphism of H defined by

$$z \longmapsto -\bar{z} \quad \text{for } z \in H$$

also induces an involutive automorphism F_∞ of $\mathbb{C}$. The induced automorphism on the cohomology group $H^1(C,\mathbb{Q})$ by w_D and F_∞ are denoted by w_D^* and F_∞^*, respectively. It is easy to check that the action of the Hecke operators on the cohomology group $H^1(C,\mathbb{Q})$ commutes with F_∞^*.

Let w_0 be an element of $\Gamma_0(D)$, which does not belong to $\Gamma_0^{\varepsilon}(D)$. Then w_0 normalizes $\Gamma_0^{\varepsilon}(D)$, and induces an involutive automorphism of C, which we denote by the same symbol w_0. The action of the Hecke algebra H^{el} on $H^1(C,\mathbb{Q})$ commutes with w_0. And w_D^* and F_∞^* also commute with w_0. Therefore, if we put

$$H^1(C,\mathbb{Q})^{Neb} = \{\delta \in H^1(C,\mathbb{Q}) \mid w_0\delta = -\delta\},$$

then by restriction H^{el}, w_0^* and F^* act on $H^1(C,\mathbb{Q})^{Neb}$.

We can restrict the intersection form of $H^1(C,\mathbb{Q})$ to $H^1(C,\mathbb{Q})^{Neb}$. Denote this by ψ_C. Then the Hecke operator T_n has the following property with respect to ψ_C:

$$\psi_C(T_n\delta, \delta') = \varepsilon_D(n)\psi_C(\delta, T_n\delta')$$

for any $\delta, \delta' \in H^1(C,\mathbb{Q})^{Neb}$.
In fact, for any cusp forms f, g of $S_2(\Gamma_0(D), \varepsilon_D)$, it is known that

$$\int_{\Gamma_0^{\varepsilon}(D)\backslash H} \{T_n f(z)\}\overline{g(z)}dxdy = \varepsilon_D(n)\int_{\Gamma_0^{\varepsilon}(D)\backslash H} f(z)\overline{\{T_n g(z)\}}dxdy,$$

where $z = x + \sqrt{-1}y \in H$ (cf. Chapter 3 of [43]).
Hence, by using the Hodge decomposition

$$H^1(C,\mathbb{Q})^{Neb} \otimes_{\mathbb{Q}} \mathbb{C} = \{2\pi i f(z)dz \mid f \in S_2(\Gamma_0(D), \varepsilon_D)\} \oplus \{\overline{2\pi i f(z)dz} \mid f \in S_2(\Gamma_0(D), \varepsilon_D)\},$$

we have

$$(\psi_C \otimes_{\mathbb{Q}} \mathbb{C})(T_n\delta, \delta') = \varepsilon_D(n)(\psi_C \otimes_{\mathbb{Q}} \mathbb{C})(\delta, T_n\delta')$$

for any $\delta, \delta' \in H^1(C,\mathbb{Q})^{Neb} \otimes_{\mathbb{Q}} \mathbb{C}$.

14.2. Let h be a normalized primitive forms of $S_2(\Gamma_0(D), \varepsilon_D)$, and let K_h be the subfield of $\mathbb{C}$ generated over $\mathbb{Q}$ by all eigenvalues a_n of Hecke operators: $T_n h = a_n h$. Let $\{\sigma_1, \ldots, \sigma_d\}$ $(d=[K_h:\mathbb{Q}])$ be the set of all embeddings of the algebraic number field K_h into $\mathbb{C}$. For each embedding $\sigma_i: K_h \hookrightarrow \mathbb{C}$, we define a function $h_i = h^{\sigma_i}$ $(1 \leqq i \leqq d)$ by

$$h_i(z) = \sum_{n=1}^{\infty} \sigma_i(a_n)\exp(2\pi i n z)$$

for the normalized primitive form h(z) with Fourier expansion

$$h(z) = \sum_{n=1}^{\infty} a_n \exp(2\pi i n z) \qquad (a_1 = 1, \quad a_n \in K_h).$$

Then, since $S_2(\Gamma_0(D), \varepsilon_D)$ has a basis consisting of modular forms whose Fourier coefficients belong to $\mathbb{Q}$, for each i the function h_i defines a primitive form of $S_2(\Gamma_0(D), \varepsilon_D)$.

For any element h of $S_2(\Gamma_0(D), \varepsilon_D)$, we put

$$\omega_h = (2\pi i)h(z)dz.$$

Then ω_h is a $\Gamma_0^{\varepsilon}(D)$-invariant 1-form on H. Hence, on passing to the quotient, it defines a holomorphic 1-form on $\Gamma_0^{\varepsilon}(D)\backslash H$. It is known that we can prolong ω_h to the compactification C of $\Gamma_0^{\varepsilon}(D)\backslash H$. Thus ω_h defines an element of $H^1_{DR}(C,\mathbb{C})^{Neb} = H^1(C,\mathbb{Q})^{Neb} \otimes_{\mathbb{Q}} \mathbb{C}$.

Let $\mathcal{H}_0^{el}$ be the subalgebra of the endomorphism ring $\mathrm{End}(H^1(C,\mathbb{Q})^{Neb})$ of the Hodge structure $H^1(C,\mathbb{Q})^{Neb}$, generated by the images of Hecke operators T_n over $\mathbb{Q}$. Since ε_D is a primitive character, $S_2(\Gamma_0(D), \varepsilon_D)$ has no old forms. Therefore, $\mathcal{H}_0^{el}$ is a commutative semisimple algebra over $\mathbb{Q}$. Hence it is a direct sum of algebraic number fields.

For any primitive form h of $S_2(\Gamma_0(D), \varepsilon_D)$, we can find a unique primitive idempotent e_h of $\mathcal{H}_0^{el}$ such that

$$\omega_h \in e_h H^1_{DR}(C,\mathbb{C})^{Neb} = e_h H^1(C,\mathbb{Q})^{Neb} \otimes_{\mathbb{Q}} \mathbb{C}.$$

Then the field

$$e_h \mathcal{H}_0^{el} e_h = e_h \mathcal{H}_0^{el} = \mathcal{H}_0^{el} e_h$$

is canonically isomorphic to K_h.

We can define a Hodge structure $H^1(M_h,\mathbb{Q})$ of weight 1 attached to the primitive form h of $S_2(\Gamma_0(D), \varepsilon_D)$ by

$$H^1(M_h,\mathbb{Q}) \underset{dfn}{=} e_h H^1(C,\mathbb{Q})^{Neb}.$$

Since $e_h \mathcal{H}_0^{el} e_h$ acts on $H^1(M_h,\mathbb{Q})$ by the restriction of the action of $\mathcal{H}_0^{el}$ on $H^1(C,\mathbb{Q})^{Neb}$ to its direct factors, via the natural isomorphism

$$e_h H_0^{el} e_h \xrightarrow{\sim} K_h$$

We can define a ring homomorphism

$$\theta_h^*: K_h \hookrightarrow \mathrm{End}(H^1(M_h, \mathbb{Q})).$$

We denote by

$$\Phi_h: H^1(M_h, \mathbb{Q}) \times H^1(M_h, \mathbb{Q}) \longrightarrow \mathbb{Q}$$

the restriction of the intersection form ψ_C^* of $H^1(C, \mathbb{Q})^{Neb}$ to the direct factor $H^1(M_h, \mathbb{Q})$. Φ_h defines a non-degenerate skewsymmetric bilinear form over $\mathbb{Q}$.

It is known that K_h is a totally imaginary quadratic extension of a totally real field k_h. We denote by ρ the non-trivial element of the Galois group of the extension K_h/k_h. Then ρ is the restriction of the complex conjugation of $\mathbb{C}$ via the natural embedding $K_h \hookrightarrow \mathbb{C}$.

Let A_h be the abelian variety attached to h with $\theta_h: K_h \hookrightarrow \mathrm{End}(A_h) \otimes_{\mathbb{Z}} \mathbb{Q}$ as in Section 7.5 of Shimura [43]. Then there is a canonical isomorphism of rational K_h-Hodge structures

$$H^1(A_h, \mathbb{Q}) \simeq H^1(M_h, \mathbb{Q}).$$

Let us consider the decomposition of $H^1(M_h, \mathbb{Q}) \otimes_{\mathbb{Q}} \mathbb{C}$ into the eigenspaces with respect to the action of K_h:

$$H^1(M_h, \mathbb{Q}) \otimes_{\mathbb{Q}} \mathbb{C} = \bigoplus_{i=1}^{d} H^1(M_h, \mathbb{Q}) \otimes_{K_h, \sigma_i} \mathbb{C}.$$

Then for each factor $H^1(M_h, \mathbb{Q}) \otimes_{K_h, \sigma_i} \mathbb{C}$, we have a Hodge decomposition

$$H^1(M_h, \mathbb{Q}) \otimes_{K_h, \sigma_i} \mathbb{C} = \mathbb{C}\omega_{h_i} \oplus \mathbb{C}\overline{\omega}_{h_i^\rho},$$

where h_i^ρ is the normalized primitive form of $S_2(\Gamma_0(D), \varepsilon_D)$, which corresponds to the embedding $\sigma_i \circ \rho: K_h \hookrightarrow \mathbb{C}$. In another way, we can define h_i^ρ by

$$h_i^\rho(z) = \overline{h_i(-\overline{z})}.$$

Put

$$H^1(M_h, \mathbb{Q})_s = \{\delta \in H^1(M_h, \mathbb{Q}) \mid F_\infty^* \delta = s\delta\},$$

where s=+, or -. Then, we can check readily that each $H^1(M_h, Q)_s$ is a K_h-module of rank 1 for each s.

For the intersection form Φ_h, we have

$$\Phi_h(\theta_h^*(a)\delta, \delta') = \Phi_h(\delta, \theta_h^*(\rho(a))\delta') \quad \text{for any } a \in K_h \text{ and any}$$

δ, δ' of $H^1(M_h, \mathbb{Q})$.

Therefore, Φ_h is written as a composition

$$\Phi_h: H^1(M_h,\mathbb{Q}) \times H^1(M_h,\mathbb{Q}) \xrightarrow[\psi_h]{} K_h \xrightarrow[tr_{K_h}/\mathbb{Q}]{} \mathbb{Q},$$

where ψ_h is a ρ-skew-Hermitian form on $H^1(M_h,\mathbb{Q})$:

$$\psi_h(\theta_h^*(a)\delta,\ \theta_h^*(b)\delta')= a\rho(b)\psi_h(\delta,\ \delta')$$

and

$$\psi_h(\delta,\ \delta')= -\rho(\psi_h(\delta',\ \delta))$$

for any a, $b\in K_h$, and any δ, $\delta'\in H^1(M_h,\mathbb{Q})$.

Let us consider the action of w_D^* on $H^1(C,\mathbb{Q})^{Neb}$. Recall that the action of the Hecke operators T_n on $H^1(C,\mathbb{Q})^{Neb}$ does not commute with w_D^*. But we have

$$w_D^* T_n w_D^* = \varepsilon_D(n)T_n \quad \text{for any } T_n$$

on $H^1(C,\mathbb{Q})^{Neb}$. Therefore, the mapping ρ_0 of $\mathrm{End}(H^1(C,\mathbb{Q})^{Neb})$ to itself

$$\rho_0 : \alpha\in H_0^{el} \longmapsto w_D^*\alpha w_D^*$$

defines an involutive automorphism of H_0^{el}. Since $w_D^*(\omega_h)$ is a constant multiple of $\omega_h\rho$ for any primitive form h of $S_2(\Gamma_0(D),\ \varepsilon_D)$,

$$w_D^*(H^1(M_h,\mathbb{Q}))= H^1(M_h,\mathbb{Q}).$$

Thus ρ_0 defines an automorphism of the field $e_h H_0^{el} e_h$, which induces the automorphism ρ of K_h via the canonical identification $e_h H_0^{el} e_h \xrightarrow{\sim} K_h$.

Let us consider the action of w_D^* on $H^1(M_h,\mathbb{Q})$, and put

$$H^1(M_h^+,\mathbb{Q})= \{\delta\in H^1(M_h,\mathbb{Q}) \mid w_D^*\delta=\delta\},$$

and

$$H^1(M_h^-,\mathbb{Q})= \{\delta\in H^1(M_h,\mathbb{Q}) \mid w_D^*\delta=-\delta\}.$$

Moreover set

$$H^1(M_h^{\pm},\mathbb{Q})_s = H^1(M_h^{\pm},\mathbb{Q})\cap H^1(M_h,\mathbb{Q})_s \quad \text{for } s=+, \text{ or } s=-.$$

14.3. <u>Lemma</u>. <u>$H^1(M_f^+,\mathbb{Q})_s$ and $H^1(M_h^-,\mathbb{Q})_s$ are k_f-modules of rank 1 for each s.</u>

<u>Proof</u>. Choose an element δ_+ of $H^1(M_h,\mathbb{Q})_+$. Then, since $w_D^*(\delta_+)$ also belongs to $H^1(M_h,\mathbb{Q})_+$ (F_∞^* and w_D^* commute), there exists an element $a\in K_h$ such that

$$w_D^*(\delta_+)= \theta_h^*(a)\delta_+.$$

Since w_D^* is involutive,

$$\delta_+=w_D^{*2}(\delta_+)=w_D^*(\theta_h^*(a)\delta_+)=\theta_h^*(\rho(a))w_D^*(\delta_+)=\theta_h^*(\rho(a))\theta_h^*(a)\delta_+.$$

Therefore $a\rho(a)=1$. Hence, Theorem 90 of Hilbert implies that there exists an element b of K_h such that $a=b/\rho(b)$. Put

$$\delta'_+=\theta_h^*(b)\delta_+.$$

Then $w_D^*(\delta'_+)=\delta'_+$. Hence $H^1(M_h^+,\mathbb{Q})_+\neq 0$. Let δ''_+ be another element of $H^1(M_h^+,\mathbb{Q})_+$. Since $H^1(M_h,\mathbb{Q})_+$ is a K_h-module of rank 1, there exists an element a' of K_h such that $\delta''_+=\theta_h^*(a')\delta'_+$. Since $w_D^*(\delta''_+)=\delta''_+$, we have $\theta_h^*(a)=\theta_h^*(\rho(a))$. Therefore $a=\rho(a)\in k_h$. Discussing similarly for $H^1(M_h^+,\mathbb{Q})_-$, we can complete a proof of our lemma. q.e.d.

14.4. Recall an abelian subvariety B_h of A_h defined by

$$B_h=(1+w_D)A_h$$

in Chapter 7 of [43], which has a homomorphism $k_h \longrightarrow \mathrm{End}(B_h)\otimes_{\mathbb{Z}}\mathbb{Q}$. Then we have a canonical isomorphism

$$H^1(B_h,\mathbb{Q})\cong H^1(M_h^+,\mathbb{Q})$$

of rational k_h-Hodge structures. Moreover the restriction Φ_h^+ of the intersection form Φ_h on $H^1(M_h,\mathbb{Q})$ to $H^1(M_h^+,\mathbb{Q})$ defines a skew-symmetric bilinear form, which coincides with the Riemann form of B_h on $H^1(B_h,\mathbb{Q})$ up to rational multiple.

The spaces

$$H^1(M_h^+,\mathbb{Q})_s \quad (s=\pm)$$

are maximal totally isotropic subspaces with respect to Φ_h^+, because F_∞ changes the orientation of $H^2(C,\mathbb{Q})$ and

$$H^1(M_h^+,\mathbb{Q}) = H^1(M_h^+,\mathbb{Q})_+\oplus H^1(M_h^+,\mathbb{Q})_-.$$

From Lemma 14.3, we can easily show that Φ_h^+ splits into a composition

$$H^1(M_h^+,\mathbb{Q})\times H^1(M_h^+,\mathbb{Q}) \xrightarrow[\psi_h^+]{} k_h \xrightarrow[\mathrm{tr}]{} \mathbb{Q},$$

where ψ_h^+ is a k_h-bilinear skewsymmetric form, and tr is the trace mapping with respect to the extension $k_h/\mathbb{Q}$.

The Hodge decomposition of

$$H^1(B_h,\mathbb{Q})\cong H^1(M_h^+,\mathbb{Q})=\bigoplus_{j=1}^{d} H^1(M_h^+,\mathbb{Q})\otimes_{k_h,\sigma_i}\mathbb{C}$$

is given by

$$H^1(M_h^+,\mathbb{Q})\otimes_{k_h,\sigma_i}\mathbb{C}=H^{1,0}\oplus H^{0,1},$$

with

$$H^{1,0}=\mathbb{C}(\omega_h+w_D^*(\omega_h)),\text{ and } H^{0,1}=\mathbb{C}(\overline{\omega}_h+w_D^*(\overline{\omega}_h)).$$

By means of Multiplicity One Theorem (Theorem 2.2), there exists a constant $c(h^\rho)$ such that

$$w_D^*(\omega_h) = c(h^\rho)\omega_{h^\rho}.$$

Since w_D^* is involutive, $c(h)c(h^\rho)=1$. As noted in [48] ($w_D \in \mathrm{Aut}(C)$)

$$\int_C \omega_h \wedge \overline{\omega}_h = \int_C \omega_{h^\rho} \wedge \overline{\omega}_{h^\rho}$$

implies that $|c(h)|^2=1$. Therefore $c(h^\rho)=\overline{c(h)}$, and $|c(h)|=1$.

Let us consider the homology group $H_1(M_h^+,\mathbb{Q})$ defined similarly as the cohomology group $H^1(M_h^+,\mathbb{Q})$. Choose two elements

$$\gamma_+ \in H_1(M_h^+,\mathbb{Q})_+ \text{ and } \gamma_- \in H_1(M_h^+,\mathbb{Q})_-$$

such that

$$\psi_{h*}^+(\gamma_+, \gamma_-) = +1,$$

where ψ_{h*}^+ is a skewsymmetric k_h-bilinear form such that

$$tr_{k_h/\mathbb{Q}}(\psi_{h*}^+)$$

is the restriction of the intersection form to $H_1(M_h^+,\mathbb{Q})$.

Form the period integrals of ω_{h_i}

$$w_+(h_i) = \frac{1}{2}\int_{\gamma_+} \{\omega_{h_i} + w_D^*(\omega_{h_i})\} = \int_{\gamma_+} \omega_{h_i}$$

and

$$w_-(h_i) = \frac{1}{2}\int_{\gamma_-} \{\omega_{h_i} + w_D^*(\omega_{h_i})\} = \int_{\gamma_-} \omega_{h_i},$$

with respect to this k_h-basis $\{\gamma_+, \gamma_-\}$ for each i ($1 \leqq i \leqq d$).
Then in terms of a basis $\{\delta_+, \delta_-\}$ of $H^1(M_h^+,\mathbb{Q})$ over k_h, corresponding to $\{\gamma_+, \gamma_-\}$ via Poincaré duality

$$H^1(M_h,\mathbb{Q}) \cong H_1(M_h,\mathbb{Q}),$$

we can write

$$\frac{1}{2}\{\omega_{h_i} + w_D^*(\omega_{h_i})\} = w_+(h_i)\delta_{-,i} - w_-(h_i)\delta_{+,i},$$

where $\delta_{+,i}$ and $\delta_{-,i}$ are the images of δ_+ and δ_- in $H^1(M_h^+,\mathbb{Q}) \otimes_{k_h,\sigma_i} \mathbb{C}$.

14.5. <u>Proposition</u>. (<u>Period relation of Riemann</u>). <u>Choose a basis $\{\gamma_+, \gamma_-\}$ of $H^1(M_h^+,\mathbb{Q})$ such that $\gamma_s \in H_1(M_h^+,\mathbb{Q})_s$ (s=+,-) and $\psi_h^+(\gamma_+, \gamma_-)=1$. And form period integrals</u>

$$w_+(h_i) = \int_{\gamma_+} \omega_{h_i}, \quad \underline{\text{and}} \quad w_-(h_i) = \int_{\gamma_-} \omega_{h_i} \quad \underline{\text{for}}\ \underline{\text{each}}\ i\ (1 \leqq i \leqq d).$$

Then,

$$\frac{1}{2}\int_C \omega_{h_i} \wedge \overline{\omega}_{h_i} = -w_+(h_i)\overline{w_-(h_i)} + \overline{w_+(h_i)}w_-(h_i)$$

$$= 2c(h_i)w_+(h_i)w_-(h_i).$$

Proof. Consider an integral

$$\int_C \frac{1}{2}\{\omega_{h_i} + w_D^*(\omega_{h_i})\} \wedge \frac{1}{2}\{\overline{\omega}_{h_i} + w_D^*(\overline{\omega}_{h_i})\}$$

$$= \frac{1}{4}\int_C (\omega_{h_i} \wedge \overline{\omega}_{h_i} + \omega_{h_i^\rho} \wedge \overline{\omega}_{h_i^\rho}) = \frac{1}{2}\int_C \omega_{h_i} \wedge \overline{\omega}_{h_i},$$

which is, on the other hand, equal to

$$-w_+(h_i)\overline{w_-(h_i)} + \overline{w_+(h_i)}w_-(h_i).$$

Since $F_\infty(\gamma_s) = s\gamma_s$ (s=+,-), we have

$$\int_{\gamma_s} \omega_h = s\int_{\gamma_s} F_\infty^*(\omega_h) = s\int_{\gamma_s} \omega_{h^\rho}.$$

Namely, $w_+(h_i^\rho) = \overline{w_+(h_i)}$, and $w_-(h_i^\rho) = -\overline{w_-(h_i)}$. Moreover, since $w_D(\gamma_s) = \gamma_s$, we have

$$w_+(h_i^\rho) = c(h_i)w_+(h_i), \quad \text{and} \quad w_-(h_i^\rho) = c(h_i)w_-(h_i).$$

Consequently,

$$\overline{w_+(h_i)} = c(h_i)w_+(h_i), \quad \text{and} \quad \overline{w_-(h_i)} = -c(h_i)w_-(h_i),$$

from which our proposition follows immediately. q.e.d.

14.6. Remark. As shown in the proof of the previous proposition,

$$\overline{w_+(h_i)} = c(h_i)w_+(h_i) \text{ and } \overline{w_-(h_i)} = -c(h_i)w_-(h_i).$$

Therefore,

$$\overline{w_-(h_i)/w_+(h_i)} = -w_-(h_i)/w_+(h_i).$$

Accordingly, the quotients $w_-(h_i)/w_+(h_i)$ ($1 \leqq i \leqq d$) are purely imaginary numbers.

Moreover it is evident from the proof of Proposition 14.5 that

$$w_-(h_i^\rho)/w_+(h_i^\rho) = w_-(h_i)/w_+(h_i)$$

for any normalized primitive form h of $S_2(\Gamma_0(D), \varepsilon_D)$.

§15.   Construction of 2-cycles.

In this section, we construct certain 2-cycles of special types on
a Hilbert modular surfaces S, and in the next section we represent the
period integrals of the primitive forms f of $S_2(SL_2(O_F))$ along these
special 2-cycles, in terms of the special values of certain types of
L-functions attached to f.

In the first four paragraphs, we discuss a special type of 2-cycles
denoted by the symbol $\gamma_{\chi,E}$, and in the later four paragraphs, another
type of cycles γ_v.

A) Construction of the 2-cycles $\gamma_{\chi,E}$ of the first type.

15.1.   Let $\mathbb{R}_+$ be the set of all real positive numbers with the usual
topology. Let us consider a product space $D_0 = \mathbb{R}_+ \times \mathbb{R}_+$, and an action
of the group E_F^2 of the squares of units of O_F defined by

$$\varepsilon((y_1, y_2)) = (\varepsilon y_1, \varepsilon' y_2) \quad \text{for } \varepsilon \in E_F^2 \text{ and for}$$

$(y_1, y_2) \in D_0$.

Let $\tilde{i}_0$ be a mapping of D_0 into $H \times H$ defined by

$$\tilde{i}_0((y_1, y_2)) = (\sqrt{-1}\, y_1, \sqrt{-1}\, y_2) \in H \times H$$

for $(y_1, y_2) \in D_0$.

Evidently, the action of E_F^2 on D_0 and the action of $SL_2(O_F)$ on $H \times H$ are
compatible with the inclusion mapping $\tilde{i}_0$. Hence, on passing to the
quotients, we have an induced mapping

$$i_0 : E_F^2 \backslash D_0 \xrightarrow{\quad\quad} S.$$

Clearly $E_F^2 \backslash D_0$ is homeomorphic to $S_1 \times \mathbb{R}$, where S_1 is 1-dimensional
sphere (=circle) and $\mathbb{R}$ is the real line. Let us compactify this
infinite cylinder by adding one point to each of two "ends" at
infinity. Then, we obtain the suspension of S_1.

Denote this compactified space by $(E_F^2 \backslash D_0)^*$, which is homeomorphic to
the 2-dimensional sphere S_2. Then, by the definition of the compactification $\overline{S}$ of S, the mapping i_0 is prolonged a continuous mapping

$$i_0^* : (E_F^2 \backslash D_0)^* \xrightarrow{\quad\quad} \overline{S}$$

of the compactifications.

Let γ_0 be the image of the fundamental class of

$$H_2((E_F^2 \backslash D_0)^*, \mathbb{Q}) \cong H_2(S_2, \mathbb{Q}) \cong \mathbb{Q}$$

in $H_2(\overline{S},\mathbb{Q})$ by means of i_0^*. Then, this cycle does not belong to the image of the canonical homomorphism

$$H_2(S,\mathbb{Q}) \longrightarrow H_2(\overline{S},\mathbb{Q}).$$

In order to construct the elements of Image$(H_2(S,\mathbb{Q}) \longrightarrow H_2(\overline{S},\mathbb{Q}))$, let us consider more general cycles of similar type.

Take an integral ideal $\mathfrak{n}=(\nu)$ ($\neq\{0\}$) of O_F. Suppose that E is a subgroup of E_F of finite index, such that for any $\varepsilon\in E$,

$$(\varepsilon-1)\in\mathfrak{n}.$$

By Dirichlet's unit theorem, such a subgroup E exists for any $\mathfrak{n}$. Let α be an element of O_F. Consider a subset $D_{\alpha/\nu}$ of $H\times H$ defined by

$$D_{\alpha/\nu}=\{(z_1,\ z_2)\in H\times H \mid \mathrm{Re}(z_1)=\alpha/\nu,\ \mathrm{Re}(z_2)=\alpha'/\nu'\ \},$$

and define an action of E on $D_{\alpha/\nu}$ by

$$\varepsilon((z_1,\ z_2))=(\varepsilon z_1-(\varepsilon-1)\tfrac{\alpha}{\nu},\ \ \varepsilon' z_2-(\varepsilon'-1)\tfrac{\alpha'}{\nu'}).$$

Note here that by assumption, we have $(\varepsilon-1)\frac{\alpha}{\nu}\in O_F$ for any $\alpha\in O_F$ and any $\varepsilon\in E$.

Consider a mapping

$$i_{\alpha/\nu,E}\colon E\backslash D_{\alpha/\nu} \longrightarrow S$$

induced from the inclusion $D_{\alpha/\nu}\hookrightarrow H\times H$ on passing to the quotients. Similarly as for D_0, we can compactify $E\backslash D_{\alpha/\nu}$ and $i_{\alpha/\nu,E}$. We denote by $\gamma_{\alpha/\nu,E}$ the image of the fundamental class of

$$H_2((E\backslash D_{\alpha/\nu})^*,\mathbb{Q}) \cong \mathbb{Q}$$

in $H_2(\overline{S},\mathbb{Q})$.

<u>Remark</u>. If we replace E by a subgroup E^d of E, consisting of d-th powers of the elements of E, we obtain d times the initial cycle. Namely we have

$$\gamma_{\alpha/\nu,E^d} = d\gamma_{\alpha/\nu,E} \quad \text{in } H_2(\overline{S},\mathbb{Q})$$

15.2. <u>Lemma</u>. <u>Let</u> $\chi\colon(O_F/\mathfrak{n})^\times \longrightarrow \mathbb{C}^\times$ <u>be a non-trivial character</u>. <u>Then an element of</u> $H_2(\overline{S},\mathbb{Q})$ <u>given by</u>

$$\sum_{\alpha \bmod \mathfrak{n}} \chi(\alpha)\gamma_{\alpha/\nu,E}$$

<u>belongs to</u> Image$(H_2(S,\mathbb{Q}) \xrightarrow[j^*]{} H_2(\overline{S},\mathbb{Q}))$.

<u>Proof</u>. Recall the long exact sequence of the relative homology groups

$$\longrightarrow H_2(S,\mathbb{Q}) \xrightarrow{\bar{j}^*} H_2(\bar{S},\mathbb{Q}) \xrightarrow{\partial} H_1(\partial S_M,\mathbb{Q}) \longrightarrow$$
$$H_2(\bar{S},\mathbb{Q}) \cong H_2(S_M \bmod \partial S_M,\mathbb{Q}) .$$

Thus it suffices to show that

$$\partial\Big(\sum_{\alpha \bmod \mathfrak{n}} \chi(\alpha)\gamma_{\alpha/\nu,E}\Big)=0.$$

It is easy to check that the image $\partial(\gamma_{\alpha/\nu,E})$ of $\gamma_{\alpha/\nu,E}$ by ∂ is given as follows.

First put

$$D_{\alpha/\nu}=\{(z_1,\ z_2)\in H\times H \mid \mathrm{Re}(z_1)=\alpha/\nu,\ \mathrm{Re}(z_2)=\alpha'/\nu',\text{ and }$$
$$\mathrm{Im}(z_1)\cdot\mathrm{Im}(z_2)=M\ \},$$

for a sufficiently large real number M.
Since

$$\mathrm{Im}\{\varepsilon z_1-(\varepsilon-1)\tfrac{\alpha}{\nu}\}\,\mathrm{Im}\{\varepsilon' z_2-(\varepsilon'-1)\tfrac{\alpha'}{\nu'}\}=\mathrm{Im}(\varepsilon z_1)\mathrm{Im}(\varepsilon' z_2)$$
$$=\mathrm{Im}(z_1)\mathrm{Im}(z_2).$$

we can restrict the action of E on $D_{\alpha/\nu}$ to $\partial D_{\alpha/\nu}$. Clearly $E\backslash\partial D_{\alpha/\nu}$ is homeomorphic to S_1, and there is a natural mapping

$$E\backslash\partial D_{\alpha/\nu} \longrightarrow \partial S_M$$

induced from the inclusion mapping $\partial D_{\alpha/\nu}\hookrightarrow(H\times H)_M$. Then the image of $\gamma_{\alpha/\nu,E}$ by ∂ is equal to the image of the fundamental class of

$$H_1(E\backslash\partial D_{\alpha/\nu},\mathbb{Q})\cong H_1(S_1,\mathbb{Q})\cong \mathbb{Q}$$

in $H_1(\partial S_M,\mathbb{Q})$ by the natural mapping $E\backslash\partial D_{\alpha/\nu} \longrightarrow \partial S_M$.

Write β for this image $\partial(\gamma_{\alpha/\nu,E})$. Then it is easy to check that β does not depend on the choice of α. Therefore,

$$\partial\Big(\sum_{\alpha \bmod \mathfrak{n}} \chi(\alpha)\gamma_{\alpha/\nu,E}\Big)=\sum_{\alpha \bmod \mathfrak{n}} \chi(\alpha)\beta=0,$$

if χ is a non-trivial character, whence the lemma follows. q.e.d.

15.3. Definition. We define a 2-cycle $\gamma_{\chi,E}$ in $H_2^{sp}(S,\mathbb{Q})$ as follows. Let χ be a Dirichlet character modulo an ideal $\mathfrak{n}$ of O_F

$$\chi:(O_F/\mathfrak{n})^\times \longrightarrow \mathbb{C}^\times,$$

and let E be a subgroup of the group E_F^2 of square units in O_F with finite index, such that $(\varepsilon-1)\in\mathfrak{n}$ for any $\varepsilon\in E$.

Then we define an element $\gamma_{\chi,E}$ of $H_2^{sp}(S,\mathbb{Q})$ as the image by the projection $\tilde{H}_2(S,\mathbb{Q}) \longrightarrow H_2^{sp}(S,\mathbb{Q})$ of the inverse image in $\tilde{H}_2(S,\mathbb{Q})$ of

$$\sum_{\alpha \bmod \mathfrak{n}} \chi(\alpha)\gamma_{\alpha/\nu,E}$$

by means of the canonical isomorphism

$$\tilde{H}_2(S,\mathbb{Q}) \underset{\text{dfn}}{=} \mathrm{Coimage}(H_2(S,\mathbb{Q}) \xrightarrow{\bar{j}^*} H_2(\bar{S},\mathbb{Q}))$$

$$\cong \mathrm{Image}(H_2(S,\mathbb{Q}) \xrightarrow{\bar{j}^*} H_2(\bar{S},\mathbb{Q})).$$

15.4. Lemma. *Let χ be a Dirichlet character $\chi:(O_F/\mathfrak{n})^\times \longrightarrow \mathbb{C}^\times$ mod $\mathfrak{n}$, and let E be an index finite subgroup of E_F^2 such that $\varepsilon-1\in\mathfrak{n}$ for all $\varepsilon\in E$. Then*

$$\gamma_{\chi,E}\in H_2^{sp}(S,\mathbb{Q})_{++} \oplus H_2^{sp}(S,\mathbb{Q})_{--}$$

or

$$\gamma_{\chi,E}\in H_2^{sp}(S,\mathbb{Q})_{+-} \oplus H_2^{sp}(S,\mathbb{Q})_{-+},$$

according as $\chi(-1)=1$ *or* $\chi(-1)=-1$.
Moreover, if $\chi(\varepsilon_0)=s1$ *and* $\chi(\varepsilon_0')=s'1$ *for the fundamental unit* ε_0 *with* $\varepsilon_0>0$ *and* $\varepsilon_0'<0$, *then*

$$\gamma_{\chi,E}\in H_2^{sp}(S,\mathbb{Q})_{ss'}.$$

Here s *and* s' *are* + *or* -.

Proof. Let $\mathfrak{n}=(\nu)$. Then for the cycle $\gamma_{\alpha/\nu,E}$ in $H_2(\bar{S},\mathbb{Q})$, we have

$$F_\infty(\gamma_{\alpha/\nu,E})= \gamma_{(-\alpha/\nu),E},\quad G_\infty(\gamma_{\alpha/\nu,E})=\gamma_{\varepsilon_0\alpha/\nu,E},$$

and

$$H_\infty(\gamma_{\alpha/\nu,E})=\gamma_{\varepsilon_0'\alpha/\nu,E}.$$

Therefore, if $\chi(-1)=+1$, then $F_\infty(\gamma_{\chi,E})=\gamma_{\chi,E}$. Hence, $\gamma_{\chi,E}$ belongs to $H_2^{sp}(S,\mathbb{Q})_{++}\oplus H_2^{sp}(S,\mathbb{Q})_{--}$.

The rest of the lemma is proved similarly. q.e.d.

B) Construction of the 2-cycles γ_v of the second type.

15.5. In the rest of this section, we construct another type of cycles which belong to the image of the canonical homomorphism

$$\bar{j}^*:H_2(S,\mathbb{Q}) \longrightarrow H_2(\bar{S},\mathbb{Q}).$$

Recall the notation of Chapter III. In the first place, for any integral vector $v\in L$ with $Q(v)=2m>0$, we define a chain $\tilde{\gamma}_v$ on S as the image by means of the natural mapping

$$j_v:\Gamma_v\backslash X_v(z_0) \longrightarrow S=\Gamma\backslash(H\times H).$$

Consider the natural compactification of the space $\Gamma_v\backslash X_v(z_0)$ (if the quotient $\Gamma_v\backslash X_v(z_0)$ itself is not compact) obtained by attaching one point to each equivalence class of cusps of the arithmetic group Γ_v. Then, it is easy to check that we can prolong j_v continuously to a mapping

$$\bar{j}_v:\overline{\Gamma_v\backslash X_v(z_0)} \longrightarrow \bar{S}.$$

We denote by Γ_v the image of the fundamental class of

$$H_2(\overline{\Gamma_v\backslash X_v(z_0)},\mathbb{Q}) \cong \mathbb{Q}$$

in $H_2(\bar{S},\mathbb{Q})$ by means of $\bar{j}_v$. The homology class γ_v does not depend on the choice of z_0, because homotopically equivalent cycles are homologous.

15.6. <u>Lemma</u>. <u>For any vector</u> $v\in L$ <u>with</u> $Q(v)>0$, <u>the cycle</u> γ_v <u>belongs to the image of the canonical homomorphism</u>

$$\bar{j}^*:H_2(S,\mathbb{Q}) \longrightarrow H_2(\bar{S},\mathbb{Q})$$

<u>induced from the inclusion</u> $\bar{j}:S\hookrightarrow\bar{S}$.

<u>Proof</u>. For each vector $v\in L$ with $Q(v)>0$, let $L_{\mathbb{Q},v}$ be the orthogonal complement of $\mathbb{Q}v$ in $L_{\mathbb{Q}}=L\otimes_{\mathbb{Z}}\mathbb{Q}$ with respect to Q. We denote by Q_v the restriction of the quadratic form Q to $L_{\mathbb{Q},v}$. Then two cases occur.

(i) <u>Anisotropic case</u>: $Q_v(x)=0$ has no solution in $L_{\mathbb{Q},v}$ except the trivial one.

In this case, there is nothing to prove, because Γ_v has no cusps and $\Gamma_v\backslash X_v(z_0)$ is compact and contained in S.

(ii) <u>Isotropic case</u>: $Q_v(x)=0$ has a non-trivial solution $x\neq 0$ in $L_{\mathbb{Q},v}$.

In this case Γ_v has cusps. The orthogonal group $SO(L_{\mathbb{Q},v};Q_v)$ over $\mathbb{Q}$ is isomorphic to $SL_2(\mathbb{Q})/\{\pm 1\}$, and of $\mathbb{Q}$-rank 1. Moreover the minimal parabolic subgroups corresponding to cusps have no reductive parts (Levi components) except ± 1.

Now let us recall the "distance" $\ell(z,\lambda)$ between the points z of $H\times H$ and cusps λ defined in Section 1.4. Put

$$(H\times H)_M=\{\, z\in H\times H \mid \ell(z,\lambda)\leqq M \text{ for all cusps } \lambda\},$$

$$\partial(H\times H)_M=\{\, z\in H\times H \mid \ell(z,\lambda)=M \text{ for some cusp } \lambda\},$$

and

$$\partial(H\times H)_{M,\infty}=\{\, z\in H\times H \mid \ell(z,\lambda)=M \,\},$$

for a real number M.

Then for a sufficiently large number M, we have

$$\partial(H \times H)_M = \bigcup_{\gamma \in \Gamma/\Gamma_\infty} \gamma(\partial(H \times H)_{M,\infty}) \qquad \text{(disjoint sum)}$$

The singular complex with rational coefficient $S((H \times H)_M)$ of $(H \times H)_M$ and its subcomplex $S(\partial(H \times H)_M)$ have natural structures of Γ-modules. Moreover as a Γ-module $S_i(\partial(H \times H)_M)$ is canonically isomorphic to an induced Γ-module of a Γ_∞-module $S_i(\partial(H \times H)_{M,\infty})$ of the i-th singular complex with rational coefficoent of $\partial(H \times H)_{M,\infty}$, with respect to the canonical inclusion $\Gamma_\infty \hookrightarrow \Gamma$ for each i (i=0,1,....). Thus Γ-complex $S(\partial(H \times H)_M)$ is a induced complex of Γ_∞-complex $S(\partial(H \times H)_{M,\infty})$.

Let $\{X_i(\Gamma), \partial_i\}_{i \in \mathbb{N}}$ (resp. $\{X_i(\Gamma_\infty), \partial_i\}_{i \in \mathbb{N}}$) be the standard complex of the discrete subgroup Γ (resp. Γ_∞) with rational coefficient. (Here $\mathbb{N}$ is the set of the natural numbers). Then for all $i \in \mathbb{N}$ we have natural monomorphisms

$$\mathbb{Q}[\Gamma/\Gamma_\infty] \otimes X_i(\Gamma_\infty) \longrightarrow X_i(\Gamma),$$

defined by

$$\gamma \otimes (\gamma_0, \gamma_1, \ldots, \gamma_i) \longmapsto (\gamma\gamma_0, \gamma\gamma_1, \ldots, \gamma\gamma_i)$$

for any homogeneous chain $(\gamma_0, \gamma_1, \ldots, \gamma_i)$ of $X_i(\Gamma_\infty)$ and any γ of a complete representative system of Γ/Γ_∞, which are compatible with the boundary operators ∂_i.

Let $\{X_i(\Gamma \bmod \Gamma_\infty), \overline{\partial}_i\}$ be the quotient complex of this monomorphism, and let $H_i(\Gamma \bmod \Gamma_\infty, \mathbb{Q})$ be the homology groups of this complex. Then we have an homomorphism of long exact sequences

$$
\begin{array}{ccccccc}
\longrightarrow & H_i(S_M, \mathbb{Q}) & \longrightarrow & H_i(S_M \bmod \partial S_M, \mathbb{Q}) & \longrightarrow & H_{i-1}(\partial S_M, \mathbb{Q}) & \longrightarrow \\
& \downarrow m_1 & \circlearrowleft & \downarrow m_2 & \circlearrowleft & \downarrow m_3 & \\
\longrightarrow & H_i(\Gamma, \mathbb{Q}) & \longrightarrow & H_i(\Gamma \bmod \Gamma_\infty, \mathbb{Q}) & \longrightarrow & H_{i-1}(\Gamma_\infty, \mathbb{Q}) & \longrightarrow
\end{array} \quad .
$$

Since for any non-trivial torsion subgroup Γ_e of $\Gamma = SL_2(O_F)/\{\pm 1\}$, $H_i(\Gamma_e, \mathbb{Q}) = 0$ (i > 0), the contractibility of $(H \times H)_M$ and $\partial(H \times H)_M$, implies that m_1 and m_3 are isomorphisms, which furthermore implies that m_2 is also an isomorphism by Five Lemma.

Let $\{c_1, \ldots, c_s\}$ be the Γ_v-equivalence classes of cusps of Γ_v, and let $\Gamma_{v,\infty,1}, \ldots, \Gamma_{v,\infty,s}$ be the the corresponding isotropy subgroups of Γ_v at representatives of these equivalence classes of cusps. Discussing similarly as above, we have a long exact sequence

$$\rightarrow H_i(\Gamma_v, \mathbb{Q}) \longrightarrow H_i(\Gamma_v \bmod \{\Gamma_{v,\infty,1}, \ldots, \Gamma_{v,\infty,s}\}, \mathbb{Q}) \longrightarrow \bigoplus_{j=1}^{s} H(\Gamma_{v,\infty,j}, \mathbb{Q}) \rightarrow$$

and canonical isomorphisms $H_i(\Gamma_v\backslash X_v(z_0),\mathbb{Q})= H_i(\Gamma_v,\mathbb{Q})$ and $H_i(\overline{\Gamma_v\backslash X_v(z_0)},\mathbb{Q})= H_i(\Gamma_v \bmod\{\Gamma_{v,\infty,1},\dots, \Gamma_{v,\infty,s}\},\mathbb{Q})$.

Now in view of the isomorphism

$$H_2(\overline{S},\mathbb{Q})= H_2(S_M \bmod \partial S_M,\mathbb{Q}),$$

to prove our lemma it is sufficient to show that in a commutative diagram induced from the inclusions $\Gamma_v\hookrightarrow\Gamma$ and $\Gamma_{v,\infty,j}\hookrightarrow\Gamma_\infty$:

$$\begin{array}{ccccc} H_2(\Gamma_v,\mathbb{Q}) & \longrightarrow & H_2(\Gamma_v \bmod \{\Gamma_{v,\infty,1},\dots,\Gamma_{v,\infty,s}\},\mathbb{Q}) & \longrightarrow & \bigoplus_{j=1}^{s} H_2(\Gamma_{v,\infty,j},\mathbb{Q}) \\ \downarrow & \circlearrowleft & \downarrow & \circlearrowleft & \downarrow m_\infty \\ H_2(\Gamma,\mathbb{Q}) & \longrightarrow & H_2(\Gamma \bmod \Gamma_\infty,\mathbb{Q}) & \longrightarrow & H_2(\Gamma_\infty,\mathbb{Q}) \longrightarrow, \end{array}$$

the homomorphism m_∞ is a zero homomorphism.

Recall that Γ_∞ is an extension of groups

$$0 \longrightarrow \Gamma_\infty^{unip} \longrightarrow \Gamma_\infty \longrightarrow \Gamma_\infty^{red} \longrightarrow 0,$$

where

$$\Gamma_\infty^{unip}= \{\pm\begin{pmatrix}1 & b\\ 0 & 1\end{pmatrix} \mid b\in O_F\}.$$

It is easy to see that the homomorphism $\Gamma_\infty \longrightarrow \Gamma_\infty^{red}$ induces an isomorphism of homology groups

$$H_1(\Gamma_\infty,\mathbb{Q}) \longrightarrow H_1(\Gamma_\infty^{red},\mathbb{Q}).$$

On the other hand, since the reductive part of any cuspidal subgroup of $SO(L_{\mathbb{Q},v};Q_v)$ is trivial, the homomorphism

$$\Gamma_{v,\infty,j} \longrightarrow \Gamma_\infty$$

factors through the unipotent part Γ_∞^{unip} of Γ_∞. Therefore the composition

$$\Gamma_{v,\infty,j} \longrightarrow \Gamma_\infty \longrightarrow \Gamma_\infty^{red}$$

is a trivial homomorphism. Hence

$$H_1(\Gamma_{v,\infty,j},\mathbb{Q}) \longrightarrow H_1(\Gamma_\infty,\mathbb{Q}) \longrightarrow H_1(\Gamma_\infty^{red},\mathbb{Q})$$

is a zero homomorphism, which proves our lemma. q.e.d.

15.7. <u>Definition</u>. For any positive integer m, we define a cycle γ_m of $\mathrm{Image}(H_2(S,\mathbb{Q}) \longrightarrow H_2(\overline{S},\mathbb{Q}))$ by a sum of cycles

$$\gamma_m= \sum_{\substack{\{v\}\\ Q(v)=2m}} \gamma_v,$$

where the summation is taken over a complete representative system of

$G_{\mathbb{Z}}$-equivalence classes of vectors v satisfying Q(v)=2m. Via the canonical identification

$$\mathrm{Coimage}(H_2(S,\mathbb{Q}) \longrightarrow H_2(\overline{S},\mathbb{Q})) \cong \mathrm{Image}(H_2(S,\mathbb{Q}) \longrightarrow H_2(\overline{S},\mathbb{Q})),$$

we denote the correponding element to γ_m in $\mathrm{Coimage}(H_2(S,\mathbb{Q}) \longrightarrow H_2(\overline{S},\mathbb{Q}))$ by the same symbol γ_m.

15.8. Lemma. For any positive integer m, $F_\infty(\gamma_m) = -\gamma_m$.

Proof. F_∞ is induced from an isomorphism of $SL_2(\mathbb{R}) \times SL_2(\mathbb{R})$ given by

$$g=(g_1,g_2) \longmapsto g'=(g_1',g_2')$$

with

$$g_i' = \begin{pmatrix} 1 & 0 \\ 0 & -1 \end{pmatrix} g_i \begin{pmatrix} 1 & 0 \\ 0 & -1 \end{pmatrix} \quad (i=1,2).$$

Therefore γ_v is mapped to $\gamma_{v'}$ by F_∞, where

$$v' = \begin{pmatrix} -a & \lambda' \\ \lambda & -b \end{pmatrix} \quad \text{for } v = \begin{pmatrix} a & \lambda' \\ \lambda & b \end{pmatrix} \in L_{\mathbb{Z}}.$$

Hence if Q(v)=2m, then Q(v')=2m. Therefore as subsets of S, we regard

$$F_\infty\Big(\bigcup_{\substack{\{v\} \\ Q(v)=2m}} \gamma_v \Big) = \bigcup_{\substack{\{v\} \\ Q(v)=2m}} \gamma_v .$$

Thus it suffices to check the change of the orientation. The orientation of $X_v(z_0)$ are $G_{\mathbb{R}}$-equivalent. Because any vector $v \in L_{\mathbb{R}}$ with $Q(v) > 0$ is $G_{\mathbb{R}}$-equivalent to

$$r\begin{pmatrix} 0 & \varepsilon_0'/\sqrt{D} \\ \varepsilon_0/\sqrt{D} & 0 \end{pmatrix} \in L_{\mathbb{R}},$$

it suffices to check the change of the orientation for $X_{v_1}(z_0)$ with

$$v_1 = \begin{pmatrix} 0 & \varepsilon_0'/\sqrt{D} \\ \varepsilon_0/\sqrt{D} & 0 \end{pmatrix}.$$

Put $z_0 = (\varepsilon_0 i, -\varepsilon_0' i)$, then $X_{v_1}(z_0)$ is given by

$$X_{v_1} = \{ (\varepsilon_0 z,\ \varepsilon_0' \overline{z}) \mid z \in H \}.$$

Clearly F_∞ mapps the point $(\varepsilon_0 z,\ \varepsilon_0'\overline{z})$ to $(\varepsilon_0(-\overline{z}),\ \varepsilon_0'(-z))$, changing the orientation ($F_\infty^*(dz \wedge \overline{dz}) = -\,dz \wedge \overline{dz}$). q.e.d.

§16. Arithmetic index theorems.

We consider two types of L-functions $L^{(1)}(s,f/F,\chi)$ and $L^{(2)}(s,f/\mathbb{Q})$ attached to each primitive form f of $S_k(SL_2(O_F))$ in this section. When k=2, we see that the special values

$$L^{(1)}(1,f/F,\chi)$$

and the residues of the poles of

$$L^{(2)}(s,f/\mathbb{Q}) \quad \text{at } s=2$$

are both represented by the period integrals of the holomorphic 2-form ω_f over the surface S. In the first three paragraphs, we discuss the first case a): $L^{(1)}(s,f/F,\chi)$, and in the rest of this section, the other case b): $L^{(2)}(s,f/\mathbb{Q})$. The main result of the case a) is Theorem 16.3.

Throughout this section, we use the convention that for any subset S of F, we denote by S_+ the set of totally positive elements of S.

Case a). The values of $L^{(1)}(s,f/F,\chi)$ at s=1 for f $S_2(SL_2(O_F)$ $(SL_2(F)$ case).

16.1. Let us recall the Mellin transformation of Hilbert modular forms (cf. Asai [2] for example). Note that $E_{F,+}=E_F^2$ in our case for the unit group E_F of O_F, by our assumption.

Suppose that

$$f(z_1, z_2)=\sum_{\mu\in O_{F,+}} C((\mu))\exp[2\pi i(\mu\omega z_1+\mu'\omega' z_2)]$$

$$=\sum_{\mu\in O_{F,+}/E_{F,+}} C((\mu))\sum_{\varepsilon\in E_{F,+}} \exp[2\pi i(\varepsilon\mu\omega z_1+\varepsilon'\mu'\omega' z_2)]$$

be the Fourier expansion of a Hilbert modular cusp form f of weight k, where ω is a totally positive element of F which generates the codifferent $\delta_F^{-1}=(\sqrt{D})^{-1}$. Note here that the Fourier coefficients $C((\mu))$ depend only on the ideal class (μ).

Put

$$L^{(1)}(s,f/F)=\sum_{\mathfrak{a}} C(\mathfrak{a})N_{F/\mathbb{Q}}(\mathfrak{a})^{-s}$$

for $s\in C$ with $\mathrm{Re}(s)>>0$, and

$$\Lambda^{(1)}(s,f/F)=D^s(2\pi)^{-2s}\Gamma(s)^2L^{(1)}(s,f/F),$$

where $\mathfrak{a}$ runs over the set of all integral ideals of O_F.

Then by an integral expression

$$L^{(1)}(s,f/F)=\int_{E_{F,+}\backslash(\mathbb{R}_+\times\mathbb{R}_+)} f(iy_1,\ iy_2)(y_1y_2)^{s-1}dy_1dy_2$$

which is valid for $\mathrm{Re}(s) > \frac{k}{2}+1$, we can show that $L^{(1)}(s,f/F)$ converges absolutely for $\mathrm{Re}(s) > \frac{k}{2}+1$, and that $\Lambda^{(1)}(s,f/F)$ is continued holomorphically to the whole complex s-plane, satisfying a functional equation

$$\Lambda^{(1)}(s,f/F)=\Lambda^{(1)}(k-s,f/F).$$

It is also known that if f is an eigenfunction of all Hecke operators $T(\mathfrak{a})$, and if it is normalized such that $C(O_F)=1$, then the eigenvalue of each $T(\mathfrak{a})$ coincides with $C(\mathfrak{a})$, and consequently we have an Euler product

$$L^{(1)}(s,f/F)=\prod_{\mathfrak{p};\text{prime}} P_{\mathfrak{p}}(s)^{-1}$$

with

$$P_{\mathfrak{p}}(s)=1-C(\mathfrak{p})N_{F/\mathbb{Q}}(\mathfrak{p})^{-s}+N_{F/\mathbb{Q}}(\mathfrak{p})^{k-1-2s},$$

where $\mathfrak{p}$ ranges over all prime ideals of O_F.

16.2. Let $\mathfrak{n}=(\nu)$ be an integral ideal of O_F, and suppose that

$$\chi:(O_F/\mathfrak{n})^\times \longrightarrow \mathbb{C}^\times$$

is a Dirichlet character mod $\mathfrak{n}$. As usual we put $\chi(\alpha)=0$, if (α) and $\mathfrak{n}$ are not coprime. Assume that χ is a primitive character, and let us consider a twisted modular form

$$f_\chi(z_1,\ z_2)=\sum_{\mu\in O_{F,+}} C((\mu))\chi(\mu)\exp[2\pi i(\mu\omega z_1+\mu'\omega' z_2)].$$

Since we have an identity

$$\sum_{\mu \bmod (\nu)} \chi(\beta)\exp[2\pi i(\frac{\alpha\beta\omega}{\nu}+\frac{\alpha'\beta'\omega'}{\nu'})]=\overline{\chi(\alpha)}G(\chi)$$

with Gaussisn sum $G(\chi)$ given by

$$G(\chi)=\sum_{\beta \bmod (\nu)} \chi(\beta)\exp[2\pi i(\frac{\beta\omega}{\nu}+\frac{\beta'\omega'}{\nu'})],$$

$f_\chi(z_1,\ z_2)$ is written as a sum

$$f_\chi(z_1,\ z_2)=\frac{1}{G(\overline{\chi})}\{\sum_{\beta \bmod (\nu)} \overline{\chi(\beta)}f(z_1+\frac{\beta}{\nu},\ z_2+\frac{\beta'}{\nu'})\}$$

Let us define the twisted Dirichlet series $L^{(1)}(s,f/F,\chi)$ by

$$L^{(1)}(s,f/F,\chi)=\sum_{\mathfrak{a}} C(\mathfrak{a})\chi(\mathfrak{a})N_{F/\mathbb{Q}}(\mathfrak{a})^{-s}.$$

Then $\Gamma(s)^2 L^{(1)}(s,f/F,\chi)$ is also continued holomorphically to the whole s-plane, and has a functional equation.

When k=2, we have the following expression of $L^{(1)}(1,f/F,\chi)$ as a period integral of ω_f.

16.3. Theorem. Let f be a primitive form of $S_2(SL_2(O_F))$. Then for any non-trivial primitive character χ, we have

$$L^{(1)}(1,f/F,\chi)=\pm\frac{1}{m}\frac{1}{(2\pi)^2}\frac{1}{G(\overline{\chi})}\int_{\gamma_{\overline{\chi},E^m_{F,+}}}\omega_f$$

for a sufficiently large integer m. Here $\overline{\chi}$ is the complex conjugate of the character χ, and here the sign $\pm$ before the integral depends on the choice of the orientation of the cycle $\gamma_{\overline{\chi},E^m_{F,+}}$ defined in Section 15.3.

Proof. Since

$$L^{(1)}(s,f/F,\chi)=\frac{1}{G(\overline{\chi})}\sum_{\beta \bmod (\nu)}\overline{\chi(\beta)}\int_{E_{F,+}\backslash\mathbb{R}^2_+} f\Big(\frac{\beta}{\nu}+iy_1,\ \frac{\beta'}{\nu'}+iy_2\Big)(y_1y_2)^{s-1}dy_1dy_2,$$

we have

$$L^{(1)}(1,f/F,\chi)=\frac{1}{G(\overline{\chi})}\sum_{\beta \bmod (\nu)}\overline{\chi(\beta)}\int_{E_{F,+}\backslash\mathbb{R}^2_+} f\Big(\frac{\beta}{\nu}+iy_1,\ \frac{\beta'}{\nu'}+iy_2\Big)dy_1dy_2$$

$$=\frac{1}{m}\frac{1}{G(\overline{\chi})}\sum_{\beta \bmod (\nu)}\overline{\chi(\beta)}\int_{E_{F,+}\backslash\mathbb{R}^2_+} f\Big(\frac{\beta}{\nu}+iy_1,\ \frac{\beta'}{\nu'}+iy_2\Big)dy_1dy_2.$$

Now make m sufficiently large so that for any $\varepsilon\in E^m_{F,+}$, $\varepsilon-1\in(\nu)$. Then, since $f(z_1+\alpha,\ z_2+\alpha')=f(z_1,\ z_2)$ for any $\alpha\in O_F$, we have

$$L^{(1)}(1,f/F,\chi)=\frac{1}{m}\frac{1}{G(\overline{\chi})}\sum_{\beta \bmod (\nu)}\overline{\chi(\beta)}\int_{\gamma_{\beta/\nu,E^m_{F,+}}} i^{-2}f(z_1,\ z_2)dz_1\wedge dz_2$$

$$=\frac{1}{m}\frac{1}{G(\overline{\chi})}\frac{1}{(2\pi)^2}\int_{\gamma_{\overline{\chi},E^m_{F,+}}}(2\pi i)^2 f(z_1,\ z_2)dz_1\wedge dz_2,$$

which shows our theorem. q.e.d.

Case b). The residues of the poles of $L^{(2)}(s,f/\mathbb{Q})$ at s=2 for primitive forms f of $S_2(SL_2(O_F))$ ($Res_{F/\mathbb{Q}}SL_2$ case).

In the rest of this section, we discuss another type of Dirichlet series $L^{(2)}(s,f/\mathbb{Q})$ attached to primitive forms $f\in S_k(SL_2(O_F))$, slightly revising the results of Asai [2]. Because we do not need some of the results on this type of L-function except in the remarks in Section 20 on the Tate conjecture and Hasse-Weil L-functions of Hilbert modular surfaces, these are given only the sketches of proofs. The reason why we contain these results which seem to be relatively irrelevant to the main results of this book, is that it is related with the Tate conjecture and the conjectures of Deligne [11]. For simplicity, in this section, from now on we assume that the discriminant D of F is a prime number.

16.4. Let f(z) be a primitive form of $S_k(SL_2(O_F))$ with Fourier expansion

$$f((z_1, z_2))=\sum_{\nu\in O_{F,+}} a((\nu))\exp[2\pi i(\nu\omega z_1+\nu'\omega' z_2)].$$

Following Asai [2], we put

$$L^{(2)}(s,f/\mathbb{Q})=\zeta(2(s-k+1))\sum_{m=1} a((m))m^{-s},$$

where $\zeta(s)$ is the Riemann zeta function.
Then this series converges absolutely for $Re(s)>k+1$. Further we put

$$\Lambda^{(2)}(s,f/\mathbb{Q})=D^{s/2}(2\pi)^{-2s}\Gamma(s)\Gamma(s-k+1)L^{(2)}(s,f/\mathbb{Q}).$$

The following theorem is a slightly revised version of the results of Asai [2].

16.5. Theorem. Let f be a primitive form of $S_k(SL_2(O_F))$. Then $\Lambda^{(2)}(s,f/\mathbb{Q})$ can be continued holomorphically to the whole s-plane except possible simple poles at s=k and s=k-1, and satisfies a functional equation

$$\Lambda^{(2)}(s,f/\mathbb{Q})=\Lambda^{(2)}(s,2k-1-s,f/\mathbb{Q}).$$

Moreover $\Lambda^{(2)}(s,f/\mathbb{Q})$ truely has simple poles at s=k and s=k-1, or has no poles at s=k and s=k-1, according as f is a lifting of an element of $S_k(\Gamma_0(D),\varepsilon_D)$ by the mapping $DN_{\mathbb{Q}}$, or not.
Proof. The first statement is Theorem 1 of [2]. Therefore it suffices to show the last statement on the poles. By Proposition 5 of [2], $\Lambda^{(2)}(s,f/\mathbb{Q})$ has in fact a simple pole at s=k, if $f=DN_{\mathbb{Q}}(h)$ for some

primitive form h of $S_k(\Gamma_0(D), \varepsilon_D)$. Therefore it is sufficient to show that $\Lambda^{(2)}(s,f/\mathbb{Q})$ has no poles, if the primitive form f does not belong to the image of the mapping DN_Q.

As shown in the proof of Theorem 1 of [2], the residue of the pole of $\Lambda^{(2)}(s,f/\mathbb{Q})$ at s=k is given by

$$I(f)=\int_{SL_2(\mathbb{Z})\backslash H} f(\varepsilon_0 z, \varepsilon_0' \bar{z}) y^k \frac{dxdy}{y^2} \qquad (z=x+\sqrt{-1}y).$$

Therefore it suffices to show that I(f)=0, if f is not lifited by DN_Q. There are two methods to show this fact.

The first method is to show that I(f) is a constant multiple of the first Fourier coefficient a(h) of $Sh_Q(f)$ by the following lemma, and to note that a(h)=0 and $Sh_Q(f)=0$ by Proposition 13.4.

Lemma. Assume that D is a prime number. Then any vector v of the lattice L (cf. Chapter III, for definition) with Q(v)=2 is $G_{\mathbb{Z}}$-equivalent to the vector

$$v_1 = \begin{pmatrix} 0 & \varepsilon_0'/\sqrt{D} \\ \varepsilon_0/\sqrt{D} & 0 \end{pmatrix}.$$

Especially the constant a(h) of Proposition 13.4 and 13.5 is equal to

$$r\int_{C_{v_1}} f(\varepsilon_0 z, \varepsilon_0' \bar{z}) y^{k-2} dxdy = rI(f),$$

where r is a rational constant independent of f and $h=c\cdot Sh_Q(f)$.

Proof of Lemma. This is equivalent to the fact that the curve F_1 of Hirzebruch-Zagier [17], [19], [56] has only one irreducible component. And when this is in fact the case by their result.

But we indicate here another proof, using the mass formula of Siegel [51]

$$\frac{M(Q,2)}{\mu(Q)} = \sum_{\substack{\{v\} \\ Q(v)=2}} \frac{\mu(Q,v)}{\mu(Q)},$$

where $\mu(Q)$ and $\mu(Q,v)$ are normalized volumes of $\Gamma\backslash(H\times H)$ and $\Gamma_v\backslash X_v(z_0)$, where M(Q,2) is the mass of Minkowski-Siegel. By the Eisenstein-Siegel formula, we can compute $\frac{M(Q,2)}{\mu(Q)}$ as the Fourier coefficient of an Eisenstein series of weight 2 with respect to $\Gamma_0(D)$ with multiplicator ε_D. In our case this coefficient is found to be

$$4\,L(2,\varepsilon_D)^{-1},$$

where $L(s,\varepsilon_D)$ is the Dirichlet L-function for the character $\varepsilon_D=(\frac{}{D})$.
On the other hand, for the vector v_1 we can compute the volume $\mu(Q,v_1)$ by the formulae

$$V_1=\int_{\Gamma_v\backslash X_v(z_0)}\frac{1}{2}(\eta_1-\eta_2)=\frac{1}{\pi}\int_{SL_2(\mathbb{Z})\backslash H}\frac{dxdy}{y^2}=-2\zeta(-1),$$

and

$$\mu(Q,v_1)=\rho_1^2(\det Q)^{-4/2}V_1,$$

where

$$\rho_n=\prod_{j=1}^{n}\frac{\pi^{k/2}}{\Gamma(k/2)}$$

(cf. Formula (37) of [51], p.122).

And the volume $\mu(Q)$ is given by the formulae

$$V=\frac{1}{(2\pi)^2}\int_{\Gamma\backslash(H\times H)}\frac{dx_1dy_1dx_2dy_2}{y_1^2y_2^2}=2\cdot\zeta_F(-1),$$

and

$$\mu(Q)=\rho_2^2(\det Q)^{-5/2}V,$$

where $\zeta_F(s)$ is the Dedekind zeta function of F (cf. Formula (13) of [51], p.110).
Since

$$-\zeta(-1)/\zeta_F(-1)=4\pi^2D^{-3/2}L(2,\varepsilon_D)^{-1}\quad\text{and}\quad\rho_2^2/\rho_1^2=\left\{\frac{\pi}{\Gamma(1)}\right\}^2=\pi^2,$$

we have

$$\frac{M(Q,2)}{\mu(Q)}=\frac{\mu(Q,v_1)}{\mu(Q)}=4\cdot L(2,\varepsilon_D)^{-1}.$$

Hence all vectors $v\in L$ with $Q(v)=2$ are $G_{\mathbb{Z}}$-equivalent to v_1. Thus the lemma is proved.

Remark. The Eisenstein series considered here coincides with E_1+E_2 in the proof of Theorem 1 of [19], p.104-106 (cf. the last remark of Section 13.7 of this book).

The other method to show $I(f)=0$ for a primitive form $f\notin \mathrm{Image}(DN_Q)$ is the following. Let α be an element of O_F such that $\alpha>0$ and $\alpha'<0$, which is not divisible by any rational integers except ±1. Similarly as $L^{(2)}(s,f/\mathbb{Q})$, we consider

$$L_\alpha^{(2)}(s,f/\mathbb{Q})\underset{dfn}{=}\zeta_A(2(s-k+1))\sum_{m=1}a((-\alpha'm))m^{-s},$$

where $A=-N_{F/\mathbb{Q}}\alpha$, and $\zeta_A(s)$ is given by

$$\zeta_A(s)=\prod_{p\nmid A}(1-p^{-s})^{-1}.$$

We can show that the residue of the pole at s=k of $L_\alpha^{(2)}(s,f/\mathbb{Q})$ is given by

$$I_\alpha(f)=\int_{\Gamma_0(A)\backslash H} f(\alpha z,\ \alpha'\bar{z})y^k\,\frac{dxdy}{y^2}.$$

It is easy to check that $I_\alpha(f)=I_{-\alpha'}(f)$. On the other hand, $L^{(2)}(s,f/\mathbb{Q})$ and $L_\alpha^{(2)}(s,f/\mathbb{Q})$ are the same Dirichlet series except finite number of Euler factors at primes dividing A. By a simple calculation, this fact implies that

$$I_\alpha(f)=a((-\alpha'))I(f).$$

Therefore we have $a((\alpha))I(f)=a((-\alpha'))I(f)$.

If f is not symmetric in $(z_1,\ z_2)$, then $a((\alpha))\neq a((-\alpha'))$ for some $\alpha\in O_F$. Hence $I(f)=0$, and $\Lambda^{(2)}(s,f/\mathbb{Q})$ has no poles at s=k.

Assume that f is symmetric. Then, by the results of Saito [38], any symmetric primitive form of $S_k(SL_2(O_F))$ is a lifting of an element of $S_k(\Gamma_0(D),\ \varepsilon_D)$ or $S_k(SL_2(\mathbb{Z}))$. For the latter case, Proposition 4 of Asai [2] tells that $I(f)=0$. Hence we have $I(f)=0$ for any primitive form $f\notin DN_Q(S_k(\Gamma_0(D),\ \varepsilon_D)$.

By the functional equation, we can discuss the pole at s=k-1 similarly. q.e.d.

16.6. <u>Remark</u>. Similarly as in the case of the Siegel modular cusp forms belonging to the Maass space (cf. Remark 5.3 of [33]), in our case, the residues of the poles of the L-functions $L^{(2)}(s,f/\mathbb{Q})$ at s=k are given by a kind of period integrals. Especially when k=2, these integrals are periods of Hilbert modular surfaces.

16.7. Let

$$f(z)=\sum_{\nu\in O_{F,+}} a((\nu))\exp[2\pi i(\nu\omega z_1+\nu'\omega' z_2)]$$

be a primitive form of $S_k(SL_2(O_F))$. Let us say that f(z) satisfies the Ramanujan-Petersson conjecture, if for any prime ideal $\mathfrak{p}\nmid(D)$ of O_F, the inequality

$$|a(\mathfrak{p})|\leq N_{F/\mathbb{Q}}(\mathfrak{p})^{(k-1)/2}$$

is valid.

16.8. <u>Theorem</u>. <u>Let</u> f <u>be</u> <u>a</u> <u>primitive</u> <u>form</u> <u>of</u> $S_k(SL_2(O_F))$, <u>which</u> <u>is</u> <u>not</u> <u>obtained</u> <u>by</u> <u>the</u> <u>lifting</u> DN_Q. <u>Assume</u> <u>that</u> f <u>satisfies</u> <u>the</u> <u>Ramanujan-Petersson</u> <u>conjecture</u>. <u>Then</u>

$$L^{(2)}(k,f/\mathbb{Q})\neq 0.$$

<u>Sketch of proof</u>. First assume that f is symmetric. Then by the result of Saito [38], f is a lifting of an element of $S_k(\Gamma_0(D),\ \varepsilon_D)$ by DN_Q, or a lifting of an element of $S_k(SL_2(\mathbb{Z}))$. By Proposition 4 of Asai [2], $L^{(2)}(k,f/\mathbb{Q})\neq 0$ for the latter case without assuming that f satisfies the Ramanujan-Petersson conjecture. Therefore, we have to check the statement of the theorem only for non-symmetric f. Then

$$\iota(f)(z_1,\ z_2)=f(z_1,\ z_2)$$

is another primitive form distinct from f.

Now let us consider a convolution of L-series

$$L^{(2)}(s,\ f\otimes\iota(f)/F) \underset{\text{dfn}}{=} \zeta_F(2(s-k+1)) \sum_{\mathfrak{a}} a(\mathfrak{a})a(\mathfrak{a}')N_{F/\mathbb{Q}}(\mathfrak{a})^{-s}$$

$(Re(s)>k+1)$, where $\mathfrak{a}$ runs over all integral ideals of O_F, and $\zeta_F(s)$ is the Dedekind zeta function. As shown in Proposition 2 of [2],

$$H^*(s)=D^{2s}(2\pi)^{-4s}\Gamma(s)^2\Gamma(s-k+1)^2L^{(2)}(s,\ f\otimes\iota(f)/F)$$

can be continued holomorphically to the whole s-plane except simple poles at s=k and s=k-1, and satisfies the functional equation

$$H^*(s)=H^*(2k-1-s).$$

The residue of the possible simple pole of $H^*(s)$ at s=k is given by

$$c\int_{SL_2(O_F)\backslash(H\times H)} f(z)\overline{\iota(f)(z)}(y_1y_2)^{k-2}dx_1dy_1dx_2dy_2,$$

where c is a constant independent of f.
Since f and $\iota(f)$ are distinct normalized primitive forms, this residue which is the Petersson inner product of f and $\iota(f)$ vanishes. Therefore $H^*(s)$ has no pole at s=k. If f satisfies the Ramanujan-Petersson conjecture, $\iota(f)$ also does so. Hence an argument similar to that of Ogg (Theorem 4 of [34]) for elliptic modular forms, implies that

$$H^*(k,f/\mathbb{Q})\neq 0.$$

Recall here the following splitting formula (Theorem 3 of [2]):

$$L^{(2)}(s,\ f\otimes\iota(f)/F)=L^{(2)}(s,f/\mathbb{Q})L^{(2)}(s,f_\chi/\mathbb{Q}),$$

where f_χ is the twisting of f with respect to the character χ of O_F induced from ε_D via $O_F/(\sqrt{D})\cong\mathbb{Z}/D\mathbb{Z}$ (cf. §3.2 of [2]), and $L^{(2)}(s,f_\chi/\mathbb{Q})$ is a L-function defined for f_χ similarly as $L^{(2)}(s,f/\mathbb{Q})$ for f.

By Theorem 16.5 and by a similar result for f_χ, we can see that $L^{(2)}(s,f/\mathbb{Q})$ and $L^{(2)}(s,f_\chi/\mathbb{Q})$ have no poles at s=k. Therefore, by the above splitting formula, $L^{(2)}(k, f\otimes\iota(f)/F)\neq 0$ implies that

$$L^{(2)}(k,f/\mathbb{Q})\neq 0 \text{ and } L^{(2)}(k,f_\chi/\mathbb{Q})\neq 0. \qquad \text{q.e.d.}$$

16.9. Remark. Assume that weight k=2. Then the assumption on the Ramanujan-Petersson conjecture is satisfied in this case for any primitive form f by the theorem of Weil-Deligne on the absolute values of the roots of congruence zeta functions, if we know that the function $L^{(2)}(s,f/\mathbb{Q})$ is a factor of the Hasse-Weil L-function of the Hilbert modular surface S defined over $\mathbb{Q}$, as claimed in § 2.3 of Casselmann [8].

$L^{(2)}(s,f/\mathbb{Q})$ has the following Euler product (cf. [2]). For any prime number $p\nmid D$, the Euler factor $L_p^{(2)}(s,f/\mathbb{Q})$ at p of $L^{(2)}(s,f/\mathbb{Q})$ is given by

$$L_p^{(2)}(s,f/\mathbb{Q})^{-1}=\begin{cases}(1-\xi_1\xi_2 p^{-s})(1-\eta_1\xi_2 p^{-s})(1-\xi_1\eta_2 p^{-s})(1-\eta_1\eta_2 p^{-s}), & \text{if } \varepsilon_D(p)=+1,\\ (1-\xi^2 p^{-2s})(1-\eta^2 p^{-2s})(1-p^2 p^{-2s}), & \text{if } \varepsilon_D(p)=-1.\end{cases}$$

Here ξ_i (i=1,2), η_i, ξ, and η are given by

$$L_p^{(1)}(s,f/F)^{-1}=\begin{cases}(1-\xi_1 p^{-s})(1-\eta_1 p^{-s})(1-\xi_2 p^{-s})(1-\eta_2 p^{-s}), & \text{if } p=\mathfrak{P}\mathfrak{P}' \text{ in } O_F,\\ (1-\xi p^{-2s})(1-\eta p^{-2s}), & \text{if (p) remains prime,}\end{cases}$$

where $L_p^{(1)}(s,f/F)$ is the Euler factor of $L^{(1)}(s,f/F)$ at p (cf.§ 3.2 and Proposition 3 of [2]).

By the theorem of Weil-Deligne, $|\xi_1\xi_2|=|\eta_1\xi_2|=|\xi_1\eta_2|=|\eta_1\eta_2|=|\xi^2|=|\eta^2|=p$, which implies the Ramanujan-Petersson conjecture for f.

§17. Period relation for the Doi-Naganuma lifting and Main Theorem B.

In this section, we discuss certain relations between the periods of the primitive form h of $S_2(\Gamma_0(D), \varepsilon_D)$ and the periods of the primitive form $f=DN_Q(h)$ of $S_2(SL_2(O_F))$, and also find that these relations imply that the abelian variety B_h attached to h (cf. Section 14) is isogenous to A_f^1 and A_f^2 (Main Theorem B).

17.1. Proposition. (Period relation for the Doi-Naganuma lifting). *Assume that the discriminant D of F is a prime number with $D\equiv 1 \bmod 4$. Let h be a normalized primitive form of $S_2(\Gamma_0(D), \varepsilon_D)$, and let $f=r_0 DN_Q(h)$ be the normalized primitive form of $S_2(SL_2(O_F))$ obtained by the lifting DN_Q from h, where r_0 is a rational constant independent of h. Let $\{\sigma_1=\mathrm{id},\dots,\sigma_d\}$ be the set of all embeddings of $k_h=K_f$ into $\mathbb{C}$, and let us denote by the same symbol σ_i, one of the two extensions of σ_i to an embedding $K_h\hookrightarrow\mathbb{C}$ for each i $(1\leqq i\leqq d)$. Then, the other extension of σ_i to $K_h\hookrightarrow\mathbb{C}$ is given by $\rho\sigma_i=\sigma_i\rho_i$, where ρ_i is the non-trivial element of $\mathrm{Gal}(\sigma_i(K_h)/\sigma_i(k_h))$ and ρ is the complex conjugation.*

For each $\sigma=\sigma_i:K_h\hookrightarrow\mathbb{C}$, define the period integrals $w_s(h_i)$ and $W_{ss'}(f_i)$ as in Theorem 4.4 and Proposition 14.5. Moreover define the constants $c(h_i)$ and $c(h_i^\rho)$ as in § 14.4.

Then we can find some non-zero elements a, b, c, and d of K_f such that

i) $W_{++}(f_i)=c(h_i)\sigma_i(a)w_+(h_i)^2$,

ii) $W_{+-}(f_i)=c(h_i)\sigma_i(b)w_+(h_i)w_-(h_i)$,

iii) $W_{-+}(f_i)=c(h_i)\sigma_i(c)w_+(h_i)w_-(h_i)$,

iv) $W_{--}(f_i)=c(h_i)\sigma_i(d)w_-(h_i)^2$,

for all embeddings $\sigma_i:K_h\hookrightarrow\mathbb{C}$ $(1\leqq i\leqq d)$.

This period realtion implies immediately the following theorem.

17.2. Main Theorem B. *Assume that D is a prime number with $D\equiv 1 \bmod 4$. Let h be a primitive form of $S_2(\Gamma_0(D), \varepsilon_D)$, and let $f=DN_Q(h)$ be its lifting in $S_2(SL_2(O_F))$. Let B_h be the Hilbert-Blumenthal abelian variety of dimension $d=[k_h:\mathbb{Q}]$ attached to h, which is defined in § 14.4. And let A_f^1 and A_f^2 be two abelian varieties defined in § 5.6, by choosing a lattice in $H^2(M_f,\mathbb{Q})$. Then $k_h=K_f$ and we have K_f-isogenies*

$$B_h \sim A_f^1 \sim A_f^2$$

<u>of Hilbert-Blumenthal abelian varieties</u> B_h, A_f^1, and A_f^2 <u>with respect to</u> K_f.

17.3. <u>Proof of Proposition</u> 17.1.

<u>Case</u> a): <u>Real periods</u>. In the first place, we discuss the realtion (i) and (iv). By Theorem 16.3, for any Dirichlet character $\chi: O_F/\mathfrak{n} \longrightarrow \mathbb{C}$ mod $\mathfrak{n}$,

$$L^{(1)}(1,f/F,\chi)=\pm\frac{1}{m}\frac{1}{(2\pi)^2}\frac{1}{G(\overline{\chi})}\int_{\gamma_{\overline{\chi},E^m_{F,+}}}\omega_f ,$$

for some $m\in\mathbb{Z}$ and some cycle $\gamma_{\overline{\chi},E^m_{F,+}} \in \tilde{H}_2(S,\mathbb{Q})$.

Now choose a Dirichlet character $\chi_0:(\mathbb{Z}/(n))^\times \longrightarrow \mathbb{C}^\times$ mod n, and put $\chi=\chi_0 N_{F/\mathbb{Q}}$, where $N_{F/\mathbb{Q}}:O_F \longrightarrow \mathbb{Z}$ is the norm mapping. Then by Lemma 15.4, we have

$$\gamma_{\chi,E}\in H_2^{sp}(S,\mathbb{Q})_{++} \quad \text{or} \quad \gamma_{\chi,E}\in H_2^{sp}(S,\mathbb{Q})_{--},$$

according as $\chi_0(-1)=+1$ or $\chi_0(-1)=-1$, because

$$\chi(\varepsilon_0)=\chi(\varepsilon_0')=\chi_0(N_{F/\mathbb{Q}}\varepsilon_0)=\chi_0(N_{F/\mathbb{Q}}\varepsilon_0')=\chi_0(-1)=+1 \quad \text{or} \quad =-1.$$

In view of the direct sum decomposition $H_2^{sp}(S,\mathbb{Q})=\bigoplus_{f\in\Xi} H_2(M_f,\mathbb{Q})$ and the fact that $H_2(M_f,\mathbb{Q})_{++}$ and $H_2(M_f,\mathbb{Q})_{--}$ are K_f-modules of rank 1, we have

$$L^{(1)}(1,f_i/F,\chi_0 N_{F/\mathbb{Q}})= \pm\frac{1}{m}\frac{1}{(2\pi)^2}\frac{1}{G(\overline{\chi})}\times\begin{cases}\sigma_i(a)W_{++}(f_i), & \text{if } \chi_0(-1)=+1,\\ \sigma_i(a)W_{--}(f_i), & \text{if } \chi_0(-1)=-1,\end{cases}$$

for all i ($1\leqq i\leqq d$), with some element a of K_f.

Discussing similarly for $L(1,h,\chi_0)$, we have

$$L(1,h_i,\chi_0)= \pm\frac{1}{(2\pi)}\cdot\frac{1}{G(\overline{\chi_0})}\sigma_i(b)\times\begin{cases}w_+(h_i), \text{ if } \chi_0(-1)=+1,\\ w_-(h_i), \text{ if } \chi_0(-1)=-1,\end{cases}$$

with some $b\in K_h$ for all i ($1\leqq i\leqq d$).

Now by using the relation of L-functions (cf. § 12.5)

$$L^{(1)}(s,f/F,\chi_0 N_{F/\mathbb{Q}})=L(s,h,\chi_0)L(s,h^\rho,\chi_0)$$

for s=1, we have

$$\text{(A)}\quad \pm\frac{1}{m}\frac{1}{(2\pi)^2}\frac{1}{G(\overline{\chi})}\sigma_i(a)W_{++}(f_i)= \pm\frac{1}{(2\pi)^2}\frac{1}{G(\overline{\chi_0})^2}\sigma_i(b)\rho\sigma_i(b)w_+(h_i)w_+(h_i^\rho),$$

or

$$\text{(B)}\quad \pm\frac{1}{m}\,\frac{1}{(2\pi)^2}\,\frac{1}{G(\overline{\chi})}\,\sigma_i(a)W_{--}(f_i)=\pm\frac{1}{(2\pi)^2}\,\frac{1}{G(\overline{\chi_0})}\,\sigma_i(b)\rho\sigma_i(b)w_-(h_i)w_-(h_i^\rho),$$

where $h_i^\rho(\tau)$ is given by $h_i^\rho(\tau)=\overline{h_i(-\overline{\tau})}$ for $\tau\in H$.
Hence,

$$\sigma_i(a)W_{++}(f_i)=\pm m\{G(\overline{\chi})/G(\overline{\chi_0})^2\}\sigma_i(b\rho_h(b))c(h_i)w_+(h_i)^2,$$

or

$$\sigma_i(a)W_{--}(f_i)=\pm m\{G(\overline{\chi})/G(\overline{\chi_0})^2\}\sigma_i(b\rho_h(b))c(h_i)w_-(h_i)^2,$$

because $w_\pm(h_i^\rho)=c(h_i)w_\pm(h_i)$.

Recall here the following

<u>Proposition</u>. (<u>Theorem</u> 2 <u>of</u> <u>Shimura</u> [48]). <u>For any primitive form</u> h <u>of</u> $S_2(\Gamma_0(D),\ \varepsilon_D)$, <u>there exists a Dirichlet character</u> χ_0 <u>with</u> $\chi_0(-1)=+1$ (<u>or with</u> $\chi_0(-1)=-1$) <u>modulo</u> n, <u>such that</u> $(n,D)=1$ <u>and</u>

$$L(1,h,\chi_0)\neq 0.$$

<u>Remark</u>. The proof of Theorem 2 of [48] is given for Haupttypus modular forms. But it is clear from its proof that it is also applicable for Nebentypus elliptic modular forms.

Because $H_1(M_h,\mathbb{Q})_+$ and $H_1(M_h,\mathbb{Q})$ are K_h-modules of rank 1, we have

$$L(1,h_i,\chi_0)=\begin{cases}\sigma_i(a')w_+(h_i), & \text{if}\ \ \chi_0(-1)=+1\\[2ex] \sigma_i(a')w_-(h_i), & \text{if}\ \ \chi_0(-1)=-1\end{cases}$$

and

$$L(1,h_i^\rho,\chi_0)=\begin{cases}\rho\sigma_i(a')w_+(h_i^\rho), & \text{if}\ \ \chi_0(-1)=+1\\[2ex] \rho\sigma_i(a')w_-(h_i^\rho), & \text{if}\ \ \chi_0(-1)=-1\end{cases}$$

for all i ($1\leqq i\leqq d$) with some element a' of K_h.

By the above proposition, we can choose a Dirichlet character χ_0 so that $a'\neq 0$. Hence for all i ($1\leqq i\leqq d$), we have

$$L(1,h_i,\chi_0)\neq 0,\quad\text{and}\quad L(1,h_i^\rho,\chi_0)\neq 0$$

for some character χ_0 with $\chi_0(-1)=+1$ (or with $\chi_0(-1)=-1$). Therefore, our relation of the periods (A) or (B) is not the trivial relation 0=0.

Since $G(\chi_0 N_{F/\mathbb{Q}})=G(\chi_0)^2$, and since $b\rho_h(b)\in k_h=K_f$, we have the required relations (i) and (iv) from (A) and (B).

17.4. Proof of Proposition 17.1 (continued).

Case b): Purely imaginary periods. Now let us prove the relations (ii) and (iii). Since f is selfconjugate, ι mapps $H_2(M_f,\mathbb{Q})_{-+}$ to $H_2(M_f,\mathbb{Q})_{+-}$, and $\iota^*(\omega_f)=-\omega_f$. Hence,

$$W_{-+}(f_i)=\int_{\gamma_{-+}}\omega_{f_i}=\int_{\iota_*(\gamma_{-+})}\iota^*(\omega_{f_i})=-\int_{\theta_*(a)\gamma_{+-}}\omega_{f_i}=\theta_i(-a)W_{+-}(f_i)$$

for some $a\in K_f$. Therefore (ii) and (iii) are equivalent mutually. Thus we have to prove that for any element $\gamma_-\in H_2(M_f,\mathbb{Q})_{+-}\oplus H_2(M_f,\mathbb{Q})_{-+}$,

$$\int_{\gamma_-}\omega_{f_i}=\sigma_i(a')W_{+-}(f_i)\quad \text{for some } a'\in K_f.$$

Recall the mappings Sh_Q and DN_Q (cf. Sections 11, 12, 13):

$$S_2(\Gamma_0(D),\,\varepsilon_D)\ \underset{Sh_Q}{\overset{DN_Q}{\rightleftarrows}}\ S_2(SL_2(O_F)),$$

and their adjointness (Proposition 13.3)

$$r\int_{\Gamma_0(D)\backslash H}\omega_{Sh_Q}\wedge\overline{\omega_h}=\int_S\omega_f\wedge\overline{\omega}_{DN_Q(h)}.$$

By Proposition 13.5 and Definition 15.7, for any primitive form h of $S_2(\Gamma_0(D),\,\varepsilon_D)$,

$$Sh_Q(DN_Q(h))=\left(\int_{\gamma_1}\omega_{DN_Q(h)}\right)(h+h^\rho),\ \text{with}\ \int_{\gamma_1}\omega_{DN_Q(h)}\neq 0.$$

Putting $f=DN_Q(h)$ in the above adjointness formula, we obtain

$$r\left(\int_{\gamma_1}\omega_f\right)\left(\int_{\Gamma_0(D)\backslash H}\omega_h\wedge\overline{\omega}_h\right)=\int_S\omega_f\wedge\overline{\omega}_f,$$

since

$$\int_{\Gamma_0(D)\backslash H}\omega_h\wedge\overline{\omega}_{h^\rho}=0.$$

Here r is a rational number independent of h.

Let us apply the period relation of Riemann-Hodge (Theorem 4.4) and the period relation of Riemann (Proposition 14.5) to this formula. Then we get

$$\left(r\int_{\gamma_1}\omega_{f_i}\right)(2c(h_i)w_+(h_i)w_-(h_i))=W_{++}(f_i)W_{--}(f_i)=-W_{+-}(f_i)W_{-+}(f_i).$$

By the relation of periods (i) and (iv), which are proved in Section 17.3,

$$W_{++}(f_i)W_{--}(f_i)=\{c(h_i)\}^2\sigma_i(ad)\{w_+(h_i)w_-(h_i)\}^2.$$

Therefore

$$\int_{\gamma_1}\omega_{f_i}=c(h_i)\sigma_i(a'')w_+(h_i)w_-(h_i) \quad \text{for some } a''\in K_f.$$

By Lemma 15.8, $\gamma_1 \in H_2(S,\mathbb{Q})_{+-} \oplus H_2(S,\mathbb{Q})_{-+}$, and by Proposition 13.5,

$$\int_{\gamma_1}\omega_{f_i}\neq 0.$$

Therefore

$$\int_{\gamma_1}\omega_{f_i}=\sigma_i(b')W_{+-}(f_i)\neq 0 \quad \text{for all } i \ (1\leqq i\leqq d)$$

with some $b'\in K_f$, which shows the relations (ii) and (iii). q.e.d.

17.5. Proof of Main Theorem B. Recall that $k_h=K_f$ for $f=DN_Q(h)$. The period moduli of the isogeny classes of the Hilbert-Blumenthal abelian varieties A_f^1 and A_f^2 are represented by the $GL_2(K_f)$-orbits of

$$(W_{+-}(f_i)/W_{++}(f_i))_{1\leqq i\leqq d} \quad \text{and} \quad (W_{-+}(f_i)/W_{++}(f_i))_{1\leqq i\leqq d}$$

in $X=(\mathbb{C}-\mathbb{R})^d$ by Theorem 6.6. By Proposition 17.1, these are equal to the $GL_2(K_f)$-orbits of

$$(\sigma_i(b/a)w_-(h_i)/w_+(h_i))_{1\leqq i\leqq d} \quad \text{and} \quad (\sigma_i(c/a)w_-(h_i)/w_+(h_i))_{1\leqq i\leqq d}$$

with b/a and $c/d\in K_f$.

On the other hand, the period modulus of the k_h-isogeny class of the Hilbert-Blumenthal abelian variety B_h is given by

$$(w_-(h_i)/w_+(h_i))_{1\leqq i\leqq d}\ .$$

Therefore, B_h and A_f^1, or A_f^2 are K_f-isogenous. q.e.d.

§18. Selfconjugate forms and transcendental cycles.

As an immediate consequence of our Main Theorems A and B, and the results of Shimura [45], we have the following theorem.

18.1. Main Theorem C. *Assume that the discriminant D of the real quadratic field F is a prime number congruent to 1 modulo 4, and that the class number of F is 1. Suppose that f is a selfconjugate form of* $S_2(SL_2(O_F))$. *Then the Lefschetz number* $\lambda(M_f)$ *of the Hodge structure* $H^2(M_f,\mathbb{Q})$ *is equal to* $3[K_f:\mathbb{Q}]$.

Proof. Since $S_2(SL_2(\mathbb{Z}))=0$, the selfconjugate forms in $S_2(SL_2(O_F))$ are obtained by the lifting DN_Q (cf. Saito [38], Zagier [56]). Therefore by Corollary 2 of Main Theorem A (§ 7.7) and Main Theorem B (§ 17.2), it suffices to show that

$$\mathrm{rank}_{k_h}\{\mathrm{End}(B_h)\otimes_{\mathbb{Z}}\mathbb{Q}\}\neq 2,$$

i.e. B_h is not of CM-type.

On the other hand, by Proposition 1.6 of [45] and by the results of [44], the jacobian variety of the modular curve $\Gamma_1(D)\backslash H^*$ has no abelian subvarieties of CM-type for a prime D (cf. Remark 1.7 of Shimura [45]), a fortiori B_h is not of CM-type. q.e.d.

18.2. Corollary of Main Theorem C. *Assume that D is a prime number congruent to 1 modulo 4, and that the class number of F is one. Assume moreover that all elements of* $S_2(SL_2(O_F))$ *are selfconjugate. Then the the Lefschetz number* $\lambda(S^*)$ *of a proper smooth model* S^* *of the Hilbert modular surface S, is equal to* $3p_g$. *Here* $p_g=\dim_{\mathbb{C}} S_2(SL_2(O_F))$ *is the geometric genus of the surface* S^*.

Proof. This corollary follows immediately from Main Theorem C, because

$$\lambda(S^*)=\lambda(W_2H^2(S,\mathbb{Q}))=\lambda(H^2_{sp}(S,\mathbb{Q}))=\sum_{f\in\Xi}\lambda(M_f)=\sum_{f\in\Xi}3[K_f:\mathbb{Q}]=3p_g.$$

q.e.d.

18.3. We discuss an application of Theorem C in this paragraph. Let f be a selfconjugate primitive form of $S_2(SL_2(O_F))$, and let us consider the antisymmetric part $H_2(M_f,\mathbb{Q})^{asym}$ of $H_2(M_f,\mathbb{Q})$ discussed in § 8, and the restriction ψ_f^a of the K_f-bilinear form ψ_f to $H_2(M_f,\mathbb{Q})^{asym}$. Then we have the following

18.4. Proposition. *Assume that the discriminant D of F is a prime congruent to 1 mod 4. Let* $\det(\psi_f^a)$ *be the determinant of the symmetric*

K_f-bilinear form

$$\psi_f^a : H_2(M_f,\mathbb{Q})^{asym} \times H_2(M_f,\mathbb{Q})^{asym} \longrightarrow K_f.$$

Then $-\det(\psi_f^a)$ *is a square of an element of* $K_f^\times$. *And consequently, the "self-intersection"* $\psi_f(\gamma,\gamma)$ *of any element* γ *of* $H_2(M_f,\mathbb{Q})^{sym}$ *is* $-a^2$ *for some* $a \in K_f$.

Proof. Assume that $-\det(\psi_f^a) \notin (K_f^\times)^2$. Then for the inverse form ψ_f^{*a} of ψ_f^a on $H^2(M_f,\mathbb{Q})^{asym}$, we have $-\det(\psi_f^{*a}) \notin (K_f^\times)^2$. Then the even Clifford algebra $C^+(H^2(M_f,\mathbb{Q})^{asym}, \psi_f^{*a})$ is a totally indefinte division quaternion algebra over K_f. Let us denote by A^{as} the abelian variety obtained from the datum $(H^2(M_f,\mathbb{Q})^{asym}(1), \psi_f^{*a})$ by using this Clifford algebra as in Section 5. Then A^{as} is of dimension $2[K_f:\mathbb{Q}]$, and there is a homomorphism

$$C^+(H^2(M_f,\mathbb{Q})^{asym}) \longrightarrow \mathrm{End}(A^{as})\otimes_{\mathbb{Z}}\mathbb{Q}.$$

Lemma. A^{as} *is isogenous to* $A_f^1 \times A_f^1$.

Proof of Lemma. Let

$$C(H^2(M_f,\mathbb{Q}))=C^+(H^2(M_f,\mathbb{Q})) \oplus C^-(H^2(M_f,\mathbb{Q}))$$

be the decomposition of the Clifford algebra $C(H^2(M_f,\mathbb{Q}))$ into the even part and the odd part. Then we have an isomorphism

$$p : C^+(H^2(M_f,\mathbb{Q})) \xrightarrow{\sim} C^-(H^2(M_f,\mathbb{Q}))$$

as left $C^+(H^2(M_f,\mathbb{Q}))$-modules. Let

$$i : C^+(H^2(M_f,\mathbb{Q})^{asym}) \hookrightarrow C^+(H^2(M_f,\mathbb{Q}))$$

be an injective homomorphism induced from the inclusion mapping

$$H^2(M_f,\mathbb{Q})^{asym} \hookrightarrow H^2(M_f,\mathbb{Q}),$$

and let

$$j : C^+(H^2(M_f,\mathbb{Q})^{asym})\delta^s \hookrightarrow C^-(H^2(M_f,\mathbb{Q}))$$

be also the natural inclusion mapping, where δ^s is a non-zero element of $H^2(M_f,\mathbb{Q})^{sym}$. Clearly, $i(a)j(b)=j(ab)$ for any a $C^+(H^2(M_f,\mathbb{Q})^{asym})$ and any b $C^+(H^2(M_f,\mathbb{Q})^{asym})\delta^s$. It is easy to check that for the complex structures J_1 and J_2 of $C^+(H^2(M_f,\mathbb{Q})^{asym})\otimes_{\mathbb{Q}}\mathbb{R}$ and $C^+(H^2(M_f,\mathbb{Q})\otimes_{\mathbb{Q}}\mathbb{R}$ defined as in § 5.7, we have $i(J_1)=p(J_2)$. Thus the composition of j and p^{-1} induces an homomorphism of abelian varieties

$$A^{as} \longrightarrow A_f^1 \times A_f^1 \times A_f^2 \times A_f^2$$

with finite kernel. Since j and p are K_f-linear, and since A_f^1 and A_f^2

are K_f-isogenous by Proposition 8.11, we have an isogeny

$$A^{as} \longrightarrow A_f^1 \times A_f^1.$$

Thus the lemma is proved.

Let us continue the proof of our proposition. By our assumption, the division quaternion $C^+(H^2(M_f,\mathbb{Q})^{asym})$ is contained in

$$M_2(\mathrm{End}_{O_f}(A_f^1)\otimes_{\mathbb{Z}}\mathbb{Q}).$$

Therefore, $\mathrm{rank}_{K_f}\{\mathrm{End}_{O_f}(A_f)\otimes_{\mathbb{Z}}\mathbb{Q}\} > 1$. Hence A_f is of CM-type by the structure theory of the endomorphism rings of abelian varieties, which contradicts the result of Main Theorem C that A_f is not of CM-type. Thus $C^+(H^2(M_f,\mathbb{Q})^{asym})$ is a matrix algebra, i.e. $-\det(\psi_f^a)\in(K_f^\times)^2$. The latter part of our proposition follows from the fact that

$$\det(\psi_f^a)\psi_f(\gamma,\gamma)\in(K_f^\times)^2 \quad \text{for any } \gamma\in H^2(M_f,\mathbb{Q})^{sym},$$

because ψ_f is a kernel form over K_f. q.e.d.

18.5. Examples. Many examples of Hilbert modular surfaces are studied by Hirzebruch and his collaborators [16], [17], [18], [19]. Especially we refer to Table 2 of Hirzebruch-Van de Ven [18].

For the real quadratic field with prime discriminant D ($\equiv 1$ mod 4), we can give here the complete list of Hilbert modular surfaces, for which the Lefschetz number λ is determined by our result (§ 18.3). In order to apply Corollary of Main Theorem C, the arithmetic genus of the symmetric Hilbert modular surface should be one. Therefore, by the results of § 5.6 of Hirzebruch [16], the discriminant of F should be $D<193$, or D=197, 229, 269, 293, or 317.

Example 1). D=29, 37, or 41: In these three cases, the Hilbert modular surface S is birational to a K3 surface (cf. [16], [18]). The geometric genus $p_g=\dim_{\mathbb{C}} S_2(SL_2(O_F))=1$. Since $\dim_{\mathbb{C}} S_2(\Gamma_0(D),\varepsilon_D)=2$, this space is spanned by a normalized primitive form h and its companion h^ρ. The Lefschetz number λ is 3 by our result. Therefore, if we write S^{min} for the proper smooth minimal model of S over $\mathbb{C}$, then the Picard number $\rho(S^{min})$ of S^{min} is given by

$$\rho(S^{min})=b_2(S^{min})-\lambda(S^{min})=22-3=19.$$

Example 2). D=229: The class number of $F=\mathbb{Q}(\sqrt{229})$ is 3. Therefore our result is not applicable to this case directly. But we indicate here that the same method is applicable mutatis mutandis.

Let $\mathfrak{a}$ be an ideal of O_F, and put

$$SL_2(O_F,\mathfrak{a})=\{\begin{pmatrix}\alpha & \beta\\ \gamma & \delta\end{pmatrix}\in SL_2(F) \mid \alpha,\ \delta\in O_F,\ \beta\in\mathfrak{a}^{-1},\ \gamma\in\mathfrak{a}\}.$$

Then for any $\mathfrak{a}$, we can find a matrix A in $GL_2^+(F)$ such that

$$A^{-1}SL_2(O_F)A=SL_2(O_F,\mathfrak{a}),$$

because the class number h of F is odd (cf. § 5.3 of Hirzebruch [16]). Therefore the surface $S_{\mathfrak{a}}= SL_2(O_F,\mathfrak{a})\backslash(H\times H)$ and S are isomorphic. Let $\{\mathfrak{a}_1,\dots,\mathfrak{a}_h\}$ be a complete set of representatives of the ideal class group of F. Then we can define a natural action of the Hecke operators $T(\mathfrak{b})$ for any integral ideal $\mathfrak{b}$ of O_F on the direct sum of the Hodge structures

$$\bigoplus_{i=1}^{h} W_2H^2(S_{\mathfrak{a}_i},\mathbb{Q}).$$

By the above isomorphism between $S_{\mathfrak{a}}$ and S, the Lefschetz number $\lambda(W_2H^2(S_{\mathfrak{a}_i},\mathbb{Q}))$ does not depend on the choice of the ideal class $\mathfrak{a}_i$. Hence,

$$\lambda(\bigoplus_i W_2H^2(S_{\mathfrak{a}_i},\mathbb{Q}))=h\cdot\lambda(W_2H^2(S,\mathbb{Q})).$$

Since the assumption h=1 is irrelevant for the theory of the lifting and L-functions, we can show Main Theorem B (§ 17.2) with adequate modifications for the case $h>1$ too. Consequently, if the assumption of Corollary of Main Theorem C (§ 18.3) is satisfied except the aasumption on the class number h, then

$$\lambda(\bigoplus W_2H^2(S_{\mathfrak{a}_i},\mathbb{Q}))=3\sum_{i=1}^{h} p_g(S_{\mathfrak{a}_i})=3h\cdot p_g(S).$$

Accordingly, $\lambda(S^*)=3\cdot p_g(S^*)$ for some (and any) proper smooth model S^* of S.

The following is the table of the Lefschetz numbers of S. The values of the gometric genera are due to Table 2 of Hirzebruch-Van de Ven [18].

D	29	37	41	53	61	73	89	97	101	109	113	137	149	157	173	181	193
p_g	1	1	1	2	2	2	3	3	4	4	4	5	6	6	7	7	9
$p_g^{s.c}$	1	1	1	2	2	2	3	3	4	4	4	5	6	6	7	7	7
λ	3	3	3	6	6	6	9	9	12	12	12	15	18	18	21	21	?

D	197	229	233	241	257	269	277	281	293	313	317
p_g	8	9	11	13	12	11	13	15	12	18	13
$p_g^{s.c}$	8	9	9	9	10	11	11	11	12	12	13
λ	24	27	?	?	?	33	?	?	36	?	39

In this table $p_g^{s.c}$ represents the dimension of the subspace of the selfconjugate forms in $S_2(SL_2(O_F))$.

Example 3). D=193, 233, 277: In these three cases, we cannot determine the Lefschetz number λ of S completely. However, in these cases, the difference $p_g - p_g^{s,c}=2$. Therefore there exist two non-selfconjugate forms in $S_2(SL_2(O_F))$. By the result of Section 9, the field of eigenvalues K_{f_i} of these primitive forms f_i (i=1,2) are quadratic extension of k_{f_i}. Since $[K_{f_i}:\mathbb{Q}]\leqq 2$, we have necessarily $K_{f_1}=K_{f_2}$, $k_{f_i}=\mathbb{Q}$, and K_{f_i} is a real quadratic extension of $\mathbb{Q}$. In these three cases, the symmetric Hilbert modular surface T =<id., ι>\S are known to be biratinally equivalent to K3 surfaces by Hirzebruch [62]. By Main Theorems A, B, and C, for the Lefschetz number λ of S, we have three possibilities: When D=193, λ=25, 27 or 29, when D=233, λ=31, 33 or 35, when D=277, λ=34, 36 or 38. Corresponding to these three possibilities of the value of λ, the Picard number $\rho(T^{min})$ of the smooth proper minimal model T^{min} of the symmetric Hilbert modular surface T is 20, 19 or 18 (cf. § 9), respectively.

Remark 1. If the first case occurs, then we can easily show that $H^2_{ét}(M_{f_1},\mathbb{Q}_\ell)$ is an abelian ℓ-adic representation of $Gal(\overline{F}/K)$, where $H^2_{ét}(M_{f_1},\mathbb{Q}_\ell)$ is the ℓ-adic cohomology group attached to f_1 defined in the next section, and where K is a finite extension over F such that a basis of the Néron-Severi group of T^{min} is defined over K.

Remark 2. If D=257, then the symmetric Hilbert modular surface T is biratinal to a K3 surface by [62]. Though the class number of F is 3, we can discuss this case similarly as the case D=193, 233 or 277.

Remark 3. As shown in [32], if the Picard number of a K3 surface X over an algebraic number field k is 19, then the Tate module $V_\ell(Br(\overline{X}))$ is isomorphic to $Sym^2(H^2(\overline{E},\mathbb{Q}_\ell))(1)$ as $Gal(\overline{\mathbb{Q}}/L)$-modlues, for some elliptic curve E defined over a finite extension L of k, or $V_\ell(Br(\overline{X}))$ is isomorphic to $V_\ell(Br(\overline{A}))$ as $Gal(\overline{\mathbb{Q}}/L)$-modules for some abelian variety A of dimension 2 defined over a finite extension L of k with a homomorphism:$B \hookrightarrow End(A)\otimes_{\mathbb{Z}}\mathbb{Q}$ of some indefinite quaternion algebra B over $\mathbb{Q}$. Here $Br(\overline{X})$ and $Br(\overline{A})$ are the Brauer groups of $\overline{X}=X\times\overline{k}$ and $\overline{A}=A\times\overline{L}$, and here Sym^2 stands for the symmetric tensor product of degree 2.

In the next section, we shall see that the first case occurs for the K3 type Hilbert modular surfaces with D=29, 37 and 41, and identify the elliptic curve E with a factor of the jacobian varieties of modular curves.

§19. Notes on ℓ-adic cohomology groups of certain Hilbert modular surfaces.

In this section, we discuss some examples on the second ℓ-adic cohomology groups attached to Hilbert modular forms of weight 2.

As a consequence of the comparison theorem of Artin [1] and the results of Rapoprt [36] on Hilbert-Blumenthal varieties, we can attach an ℓ-adic representation of $\mathrm{Gal}(\overline{F}/F)$ to each primitive forms of weight 2. By using the results of Deligne [10] on the ℓ-adic cohomology of K3 surfaces and our Main Theorem B, we can find a few examples (D=29, 37 or 41) of primitive forms such that the restrictions of the above representations of $\mathrm{Gal}(\overline{F}/F)$ attached to them to the subgroup $\mathrm{Gal}(\overline{F}/L)$ for a sufficiently large finite extension L of F, is tensor products of two ℓ-adic representations attached to the first ℓ-adic cohomology groups of ceratin elliptic curves which appear as factors of the jacobian varieties of modular curves.

19.1. In the first place, we attach an ℓ-adic representation to each primitive Hilbert modular cusp form of weight 2. By Shimura [46], or by Rapoport [36], our Hilbert modular surafce S, which is the coarse moduli scheme of Hilbert-Blumenthal abelian varieties, has a canonical model defined over $\mathbb{Q}$. Let us denote this model over $\mathbb{Q}$ by the same symbol S by an abuse of notation. Rapoport [36] construct a toroidal compactification $\tilde{S}$ of S defined over $\mathbb{Q}$, which is smooth outside of the finite number of quotient sigularities of S.

Fix a prime number ℓ, and put

$$W_2H^2_{\acute{e}t}(S\times\overline{\mathbb{Q}},\mathbb{Q}_\ell)=\mathrm{Image}(H^2_{\acute{e}t}(S\times\overline{\mathbb{Q}},\mathbb{Q}_\ell)\longrightarrow H^2_{\acute{e}t}(S\times\overline{\mathbb{Q}},\mathbb{Q}_\ell)).$$

Then by the comparison theorem of Artin [1], we have a canonical isomorphism

$$W_2H^2_{\acute{e}t}(\tilde{S}\times\mathbb{Q},\mathbb{Q}_\ell)\cong W_2H^2(S(\mathbb{C})^{an},\mathbb{Q})\otimes_{\mathbb{Q}}\mathbb{Q}_\ell,$$

where $S(\mathbb{C})$ is the $\mathbb{C}$-valued points of S.

The Galois group $\mathrm{Gal}(\overline{\mathbb{Q}}/\mathbb{Q})$, and accordingly its subgroup $\mathrm{Gal}(\overline{F}/F)$ acts on $W_2H^2_{\acute{e}t}(S\times\overline{\mathbb{Q}},\mathbb{Q}_\ell)$.

The Hecke operators $T(\mathfrak{a})$ are algebraic correspodences of the surface S: $T(\mathfrak{a})\hookrightarrow S\times S$. Both projections $T(\mathfrak{a})\longrightarrow S$ are finite flat, and étale except over finite number of points of S. Hence these correspondences $T(\mathfrak{a})$ induce an action of the Hecke algebra H on $W_2H^2_{\acute{e}t}(S\times\overline{\mathbb{Q}},\mathbb{Q}_\ell)$. By Shimura [46], these correspondences are defined over F. Hence the action of H on $W_2H^2_{et}(S\times\overline{\mathbb{Q}},\mathbb{Q}_\ell)$ commutes with that of $\mathrm{Gal}(\overline{F}/F)$.

Fix an integer $n\geqq 3$. Let S_n be the fine moduli scheme of Hilbert-Blumenthal abelian surfaces with the endomorphism ring O_F and the level n structure, and let $f_n:A_n \longrightarrow S_n$ be the universal abelian scheme over S_n. The first direct image $R^1f_{n*}O_{A_n}$ of O_{A_n} with respect to f_n, which is a $O_{S_n}\otimes_{\mathbb{Z}}O_F$-module of rank 1, is decomposed into a direct sum of two invertible sheaves $L_1\oplus L_2$ over S_n, according to the decomposition of $F\otimes_{\mathbb{Z}}O_F\cong F\oplus F$-algebra $O_{S_n}\otimes_{\mathbb{Z}}O_F$. Some multiples L_i^m of L_i ($i=1,2$) descend to invertible sheaves M_i ($i=1,2$) with respect to the base change S_n/S, respectively. These two sheaves M_1 and M_2 define two elements of $\mathrm{Pic}(S)=H^2_{\acute{e}t}(S,\mathbb{G}_m)$. The Kummer sequence defines the ℓ-adic Chen classes $c^1_\ell(M_i)$ of M_i ($i=1,2$) in $H^2_{\acute{e}t}(S\times\overline{\mathbb{Q}},\mathbb{Q}_\ell)(1)$. Since the complex Chern classes $c^1(M_i\otimes\mathbb{C})$ belong to $W_2H^2(S(\mathbb{C})^{an},\mathbb{Q})$ (cf.§1.9), the comparison theorem of Artin implies that $c^1_\ell(M_i)$ also belong to

$$[W_2H^2_{\acute{e}t}(S\times\overline{\mathbb{Q}},\mathbb{Q}_\ell)(1)]^{\mathrm{Gal}(\overline{F}/F)}.$$

Let

$$H^2_{sp}(S\times\overline{\mathbb{Q}},\mathbb{Q}_\ell)$$

be the orthogonal complement of $\mathbb{Q}_\ell c^1_\ell(M_1)+\mathbb{Q}_\ell c^1_\ell(M_2)$ with respect to the ℓ-adic intersection form on $W_2H^2_{et}(S\times\overline{\mathbb{Q}},\mathbb{Q}_\ell)$, which is non-degenerate by the comparison theorem. Then $H^2_{sp}(S\times\overline{\mathbb{Q}},\mathbb{Q}_\ell)$ is an H-submodule of $W_2H^2(S\times\overline{\mathbb{Q}},\mathbb{Q}_\ell)$. Let $H_{0,\ell}$ be the subalgebra over $\mathbb{Q}_\ell$ of $\mathrm{End}(H^2_{sp}(S\times\overline{\mathbb{Q}},\mathbb{Q}_\ell))$ generated by the images of the elements of H. Then $H_{0,\ell}\cong H_0\otimes_{\mathbb{Q}}\mathbb{Q}_\ell$ by the comparison theorem.

Let f be a primitive form of $S_2(SL_2(O_F))$, and let K_f be the field of eigenvalues. Let e_f be the primitive idempotent of H_0 corresponding to f (cf. § 2.3-2.5) . Then via the isomorphism $e_fH_0e_f=e_fH_0=H_0e_f \xrightarrow{\sim} K_f$, we have a $K_f\otimes_{\mathbb{Q}}\mathbb{Q}_\ell$-module

$$H^2_{\acute{e}t}(M_f,\mathbb{Q}_\ell)\underset{dfn}{=}e_fH^2_{sp}(S\times\overline{\mathbb{Q}},\mathbb{Q}_\ell)$$

with an action of $\mathrm{Gal}(\overline{F}/F)$. The comparison theorem implies that

$$H^2_{\acute{e}t}(M_f,\mathbb{Q}_\ell)=e_fH^2_{sp}(S\times\overline{\mathbb{Q}},\mathbb{Q}_\ell)\cong e_fH^2_{sp}(S(\mathbb{C})^{an},\mathbb{Q})\otimes\mathbb{Q}_\ell=H^2(M_f,\mathbb{Q})\otimes\mathbb{Q}_\ell,$$

accordingly $H^2_{\acute{e}t}(M_f,\mathbb{Q}_\ell)$ is a $K_f\otimes_{\mathbb{Q}}\mathbb{Q}_\ell$-module of rank 4.

19.2. Suppose that our Hilbert modular surafce S is birationally equivalent to a K3 surface. Such a surface exists. In fact, by [16] or by [18], S is birationally equivalent to a K3 surface for the real quadratic field $F=\mathbb{Q}(\sqrt{D})$ with D=29, 37 or 41.

In general, a theorem of Deligne [10] tells that for any K3 surface X over a field K, there exist a big abelian variety A over a finite extension K' of K and a monomorphism of $\mathrm{Gal}(\overline{K}/K')$-modules

$$H^2_{\acute{e}t}(X\times\overline{K},\mathbb{Q}_\ell) \hookrightarrow H^1_{\acute{e}t}(A\times\overline{K}',\mathbb{Q}_\ell)\otimes H^1_{\acute{e}t}(A\times\overline{K}',\mathbb{Q}_\ell).$$

Combining this result with Main Theorem B, we have the following

19.3. Example-Theorem. Assume that D=29, 37 or 41, and let h and h^ρ be a primitive form and its companion which spann $S_2(\Gamma_0(D),\varepsilon_D)$ (note that $\dim_{\mathbb{C}} S_2(\Gamma_0(D),\varepsilon_D)=2$ for D=29, 37 or 41). Then the abelian variety B_h is an elliptic curve defined over F, and for sufficiently large finite extension L of F, we have an isomorphism of $\mathrm{Gal}(\overline{\mathbb{Q}}/L)$-modules

$$H^2_{sp}(S\times\overline{\mathbb{Q}},\mathbb{Q}_\ell)=H^2(M_f,\mathbb{Q}_\ell)\cong H^2_{\acute{e}t}(B_h\times_F\overline{\mathbb{Q}},\mathbb{Q}_\ell)\otimes H^2_{\acute{e}t}(B_h\times_F\overline{\mathbb{Q}},\mathbb{Q}_\ell),$$

where $f=DN_Q(h)=DN_Q(h^\rho)$ is the generator of $S_2(SL_2(O_F))$.

Proof. The Hilbert modular surface S is defined over $\mathbb{Q}$. And its Satake compactification $\overline{S}$ and cusp singularities are defined over a finite extension of $\mathbb{Q}$. Therefore a desingularization $\tilde{S}$ of $\overline{S}$ is also defined over a finite extension of $\mathbb{Q}$. Since $\tilde{S}$ is birationally equivalent to a K3 surface, its exceptional curves are zero divisors of the holomorphic 2-form on $\tilde{S}$, accordingly they are defined over a finite extension of $\mathbb{Q}$. Hence the minimal model S^{min} of $\tilde{S}$ is also defined over a finite extension of $\mathbb{Q}$.

In our case $\dim_{\mathbb{C}} S_2(SL_2(O_F))=p_g=1$. Let f be the primitive form of $S_2(SL_2(O_F))$ unique up to constant multiple. Then f is selfconjugate. Let ι be the involutive automorphism of S defined in § 8. Then

$$H^2_{sp}(S\times\overline{\mathbb{Q}},\mathbb{Q}_\ell)=H^2(M_f,\mathbb{Q}_\ell)=H^2(M_f,\mathbb{Q}_\ell)^{asym}\oplus H^2(M_f,\mathbb{Q}_\ell)^{sym},$$

where $H^2(M_f,\mathbb{Q}_\ell)^{asym}=\{\delta\in H^2(M_f,\mathbb{Q}_\ell)\mid \iota^*(\delta)=-\delta\}$, and $H^2(M_f,\mathbb{Q}_\ell)^{sym}=\{\delta\in H^2(M_f,\mathbb{Q}_\ell)\mid \iota^*(\delta)=\delta\}$.

By the comparison theorem of Artin and the results of § 8, $H^2(M_f,\mathbb{Q}_\ell)^{sym}$ is a $\mathbb{Q}_\ell$-module of rank 1 generated by algebraic cycles. Since ι is defined over F, the above decomposition of $H^2(M_f,\mathbb{Q}_\ell)$ is a decomposition of $\mathrm{Gal}(\overline{F}/F)$-module.

Let $NS(S^{min}\times\overline{\mathbb{Q}})$ be the Neron-Severi group of $S^{min}\times\overline{\mathbb{Q}}$, and let

$$c^1_\ell: NS(S^{min}\times\overline{\mathbb{Q}})\otimes_{\mathbb{Z}}\mathbb{Q}_\ell \longrightarrow H^2(S^{min}\times\overline{\mathbb{Q}},\mathbb{Q}_\ell)(1)$$

be the cycle mapping. Denote by $V_\ell(S^{min})(1)$ the orthogonal complement of the image of c^1_ℓ with respect to the intersection form. Then, noting the fact that $\rho(S^{min}(C))=19$ (cf. § 18.5, Example 1), we can check that there exists an isomorphism

$$V_\ell(S^{min})(1) \xrightarrow{\sim} H^2(M_f,\mathbb{Q}_\ell)^{asym}(1)$$

of $Gal(\overline{\mathbb{Q}}/L_1)$-modules for some finite extension L_1 of F.

Let U be the sublattice of $H^2(S^{min}(\mathbb{C})^{an},\mathbb{Z})$ generated by algebraic cycles, and let V be the orthogonal complement of $U\otimes_{\mathbb{Z}}\mathbb{Q}$ in $H^2(S^{min}(\mathbb{C})^{an},\mathbb{Q})$ with respect to the intersection form. Then we have a natural isomorphism of rational Hodge structures

$$H^2(M_f,\mathbb{Q})^{asym} \cong V,$$

because $H^2(M_f,\mathbb{Q})^{sym}$ is generated by algebraic cycles and because $\rho(S^{min}(\mathbb{C}))=19$.

Now let us consider the deformations of $S^{min}\times\mathbb{C}$. Here we fix not only a polarization of $S^{min}\times\mathbb{C}$, but also an assigned sublattice U of $H^2(S^{min}(\mathbb{C})^{an},\mathbb{Z})$ corresponding to algebraic cycles, to consider the deformations of S^{min}. It is easy to check that the number of moduli of the formal universal deformation $\mathcal{S} \longrightarrow M$ is given by

$$\dim_{\mathbb{C}} H^2(S^{min}, \Omega^1_{S^{min}}) - \mathrm{rank}_{\mathbb{Z}} U = 20-19=1.$$

Applying the argument of Deligne [10] § 6, we have a family

$$f:\mathcal{S} \longrightarrow M$$

of K3 surfaces with the Picard number of each fibre $\rho(f^{-1}(m))\geqq 19$, over some scheme M, such that for some point $o\in M$, $f^{-1}(o)=S^{min}$. We can consider a variation of Hodge structures $V(f) \underset{dfn}{=} R^2f_*\mathbb{Z}/(U\times M)$ such that $V(f)_o=V$. Construct a family of abelian varieties $g:A \longrightarrow M$ of relative dimension 2, applying the formalism to use the Clifford algebras over the variation of Hodge structures in [10] (and in § 5). Similarly as Proposition 6.5 of [10], we can show that $A_o=g^{-1}(o)$ is defined over a finite extension of $\mathbb{Q}$, and we have an isomorphism

$$(\#) \quad \bigoplus_{i=0} \{ \bigwedge^{2i} V_\ell(S^{min})(1) \} \xrightarrow{\sim} \underline{\mathrm{End}}_{C^+}(H^2(A_o\times\overline{\mathbb{Q}},\mathbb{Q}_\ell))$$

of $Gal(\overline{\mathbb{Q}}/L_2)$-module for some finite extension L_2 of $\mathbb{Q}$. Here C^+ is the even Clifford algebra $C^+(V, \psi)$ with the restriction ψ to V of the intersection form, and here $\underline{\mathrm{End}}_{C^+}$ is the $Gal(\overline{\mathbb{Q}}/L_2)$-module of the commutants in $\underline{\mathrm{End}}(H^2(A\times\overline{\mathbb{Q}},\mathbb{Q}_\ell))$ of C^+.

Via the isomorphism of polarized rational Hodge structures

$$V \cong H^2(M_f,\mathbb{Q})^{asym}$$

and by Proposition 18.4, the even Clifford algebra $C^+=C^+(H^2(M_f,\mathbb{Q})^{asym})$ is isomorphic to the matrix algebra $M_2(\mathbb{Q})$ of size 2 over $\mathbb{Q}$. Hence A_o

is isogenous to a product $E \times E$ of an elliptic curve E defined over a finite extension of F. By Main Theorem B (§ 17.2), this elliptic curve E is isogenous to B_h over $\mathbb{C}$. Since both E and B_h are defined over a finite extension of F, there is an isogeny $E \longrightarrow B_h$ defined over a finite extension of F. Therefore there is an isomorphism

$$H^1(E \times \overline{\mathbb{Q}}, \mathbb{Q}_\ell) = H^1(B_h \times \overline{\mathbb{Q}}, \mathbb{Q}_\ell)$$

of $\mathrm{Gal}(\overline{\mathbb{Q}}/L_3)$-modules for a finite extension L_3 of F. By maens of the isomorphism

$$V_\ell(S^{min})(1) \xrightarrow{\sim} H^2(M_f, \mathbb{Q}_\ell)^{asym}(1)$$

of $\mathrm{Gal}(\overline{\mathbb{Q}}/L_1)$-modules, the isomorphism (#) implies the isomorphisms

$$\mathbb{Q}_\ell \oplus [\overset{2}{\wedge}\{H^2(M_f, \mathbb{Q}_\ell)^{asym}(1)\}] \xrightarrow{\sim} \underline{\mathrm{End}}(H^1(E \times \overline{\mathbb{Q}}, \mathbb{Q}_\ell))$$
$$\xrightarrow{\sim} \underline{\mathrm{End}}(H^1(B_h \times \overline{\mathbb{Q}}, \mathbb{Q}_\ell))$$

of $\mathrm{Gal}(\overline{\mathbb{Q}}/L)$-modules for a sufficiently large finite extension L of F.

Because there is an ℓ-adic polarization of $\mathrm{Gal}(\overline{\mathbb{Q}}/F)$-module

$$\psi_\ell : H^2(M_f, \mathbb{Q}_\ell)^{asym} \times H^2(M_f, \mathbb{Q}_\ell)^{asym} \longrightarrow \mathbb{Q}_\ell(-2),$$

we have isomorphisms

$$\overset{2}{\wedge}\{H^2(M_f, \mathbb{Q}_\ell)^{asym}(1)\} \xrightarrow{\sim} \{H^2(M_f, \mathbb{Q}_\ell)^{asym}(1)\}^\vee$$

and

$$\{H^2(M_f, \mathbb{Q}_\ell)^{asym}(1)\}^\vee \xrightarrow{\sim} H^2(M_f, \mathbb{Q}_\ell)^{asym}(1)$$

of $\mathrm{Gal}(\overline{\mathbb{Q}}/F)$-modules. On the other hand, since the $\mathrm{Gal}(\overline{\mathbb{Q}}/F)$-module $H^1(B_h \times \overline{\mathbb{Q}}, \mathbb{Q}_\ell)$ has an ℓ-adic polarization

$$H^1(B_h \times \overline{\mathbb{Q}}, \mathbb{Q}_\ell) \times H^1(B_h \times \overline{\mathbb{Q}}, \mathbb{Q}_\ell) \longrightarrow \mathbb{Q}_\ell(-1),$$

there is an isomorphism of $\mathrm{Gal}(\overline{\mathbb{Q}}/F)$-modules

$$\underline{\mathrm{End}}(H^1(B_h \times \overline{\mathbb{Q}}, \mathbb{Q}_\ell)) \xrightarrow{\sim} H^1(B_h \times \overline{\mathbb{Q}}, \mathbb{Q}_\ell) \times H^1(B_h \times \overline{\mathbb{Q}}, \mathbb{Q}_\ell)\ (1).$$

Since $H^2(M_f, \mathbb{Q}_\ell)^{sym}$ is generated by algebraic cycles and of rank 1, there is an isomorphism of $\mathrm{Gal}(\overline{\mathbb{Q}}/L)$-modules for some finte extension L of F:

$$H^2(M_f, \mathbb{Q}_\ell)^{sym} \xrightarrow{\sim} \mathbb{Q}_\ell(-1).$$

In view of all these isomorphisms, we have an isomorphism

$$H^2(M_f, \mathbb{Q}_\ell)(1) \xrightarrow{\sim} H^1(B_h \times \overline{\mathbb{Q}}, \mathbb{Q}_\ell) \otimes_{\mathbb{Q}_\ell} H^1(B_h \times \overline{\mathbb{Q}}, \mathbb{Q}_\ell)(1)$$

of $\mathrm{Gal}(\overline{\mathbb{Q}}/L)$-modules, whence our theorem follows. q.e.d.

Remark. We sometimes omitted the subscript "ét" in the symbols of the ℓ-adic cohomology groups in the above proof.

§20. Remarks.

20.1. Let us discuss the relation of our results with the Tate conjecture on the poles of L-functions and algebraic cycles. We assume that the discriminat D of the real quadratic field F is a prime number.

Because we are discussing non-compact surfaces, we cannot use the results of Langlands [21] as itself. But it is natural to expect that the Hasse-Weil L-function corresponding to the $\mathrm{Gal}(\overline{\mathbb{Q}}/\mathbb{Q})$-module $W_2H^2_{\acute{e}t}(S\times\overline{\mathbb{Q}},\mathbb{Q}_\ell)$ is given by

$$L(s,W_2H^2(S/\mathbb{Q})) \underset{\mathrm{dfn}}{=} \zeta_F(s-1)\prod_f L^{(2)}(s,f/\mathbb{Q}),$$

where the product with respect to f is taken over all normalized primitive forms f of $S_2(SL_2(O_F))$ (cf. Casselmann [8]). Here $L^{(2)}(s,f/\mathbb{Q})$ is the L-function defined in §16.4.

Let us assume this for a while. By Theorems 16.5, 16.8 and Remark 16.9, $L^{(2)}(s,f/\mathbb{Q})$ has a pole at s=2, if f is obtained by the Doi-Naganuma lifting DN_Q, and $L^{(2)}(s,f/\mathbb{Q})$ has neither pole nor zero at s=2, if f is not selfconjugate. Therefore the order of pole of $L(s,W_2H^2(S/\mathbb{Q}))$ at s=2 is equal to the number of normalized selfconjugate forms in $S_2(SL_2(O_F))$ plus 1.

Now note that the algebraic cycles F_N investigated by Hirzebruch-Zagier are defined over $\mathbb{Q}$, and also that the algebraic cycle represented by the Chern form $\eta_1+\eta_2$ is defined over $\mathbb{Q}$. We have determined the Picard number of special type of Hilbert modular surfaces such that all the elements of $S_2(SL_2(O_F))$ is obtained by the mapping DN_Q. We can see that for these surfaces, the Tate conjecture is valid over $\mathbb{Q}$.

Very little is known about the base change of modular forms on $\mathrm{Res}_{F/\mathbb{Q}}SL_2$ and their L-functions, so we are ignorant of the validity of the Tate conjecture over arbitrary algebraic number fields, even for these special Hilbert modular surfaces.

However, the following seems to be plausible

Hope 1. There are no algebraic cycles defined over $\mathbb{Q}$ on S except those cycles F_N constructed by Hirzebruch–Zagier.

Remark 1. When we consider the Hilbert modular surfaces corresponding the congruence subgroups of $SL_2(O_F)$ in place of the full modular group $SL_2(O_F)$, there is a reason to believe that the analogues of the curve F_N do not generate all algebraic cycles of $H^2_{sp}(S,\mathbb{Q})$ defined over $\mathbb{C}$. We shall discuss this problem elsewhere.

Let us assume that the discriminant D of F is a product of prime numbers p with $p\equiv 1 \bmod 4$. By Main Theorem A and its corollary (cf. §17), the analogue of Hope 1 over $\mathbb{C}$ is equivalent to the following.

Hope 2. *The abelian varieties A_f and A_f are K_f-isogenous for any non-selfconjugate primitive forms f in $S_2(SL_2(O_F))$, and for any selfconjugate form f of $S_2(SL_2(O_F))$, the K_f-isogenous abelian varieties A_f^1 and A_f^2 have the endomorphism rings such that* $\mathrm{rank}_{K_f}(\mathrm{End}(A_f^i)\otimes_{\mathbb{Z}}\mathbb{Q})=1$ $(i=1,2)$.

We have precise knowledge on the endomorphism rings of the abelian varieties B_h for the primitive forms h of $S_2(\Gamma_0(D),\ \varepsilon_D)$ by Ribet [37] and Momose [25]. Moreover we can prove Main Theorem B (§17) for composite discriminant D. It seems easy to settle the latter part of Hope 2 for any real quadratic field with discriminant $D=\prod_{p\equiv 1 \bmod 4} p$.

Remark 2. It seems difficult to show that $H^2(M_f,\mathbb{Q})_{alg}=\{0\}$, or equivalently to show that A_f^1 and A_f^2 are not K_f-isogenous for any non-selfconjugate primitive forms of $S_2(SL_2(O_F))$. But if f is not of level 1, i.e. if the primitive form f belongs to $S_2(\Gamma_0(\pi))$ of some congruence subgroup $\Gamma_0(\pi)$ of $SL_2(O_F)$, then there is a sorites to show that $H^2(M_f,\mathbb{Q})$ has no algebraic cycles. Details are discussed elsewhere.

20.2. Recall Main Conjecture A^{split} of Chapter 0, iv). Because our construction of the abelian varieties A_f^1 and A_f^2 is based on a transcendental method, in general we are ignorant whether our abelian varieties A_f^i $(i=1,2)$ are defined over algebraic number fields, or not. But, if f is a selfconjugate form, then by Theorem B of §17, the isogeny class $A_f^1\otimes\mathbb{Q}=A_f^2\otimes\mathbb{Q}$ contains the abelian variety B_h defined over F for the primitive form $h\in S_2(\Gamma_0(D),\ \varepsilon_D)$ satisfying $f=DN_Q(h)$ (cf. Shimura [43], Chap.7). Thus as a special case of Main Conjecture A^{split} of Chapter 0, it seems to be reasonable to expect the following.

Conjecture 1. *Let f be a primitive form of $S_2(SL_2(O_F))$. Then either of the two isogeny classes $A_f^i\otimes\mathbb{Q}$ (i=1, or 2) of abelian varieties contains a Hilbert-Blumenthal abelian variety A_f^i defined over F with $\theta^i:K_f \hookrightarrow \mathrm{End}(A_f^i)\otimes_{\mathbb{Z}}\mathbb{Q}$ over F. Moreover the conjugate of the Hilbert-Blumenthal abelian variety A_f^1 with respect to the extension F/$\mathbb{Q}$ is K_f-isogenous to A_f^2.*

Moreover the following conjecture also seems to be very plausible.

Conjecture 2. *Let f be a primitive form of $S_2(SL_2(O_F))$, let A_f^1 and A_f^2 be two abelian varieties defined over F, whose existence is expected by Conjecture 1. Then either of the Hasse-Weil L-functions $L(s,H^1(A_f^i))$ of*

A_f^i $(i=1,2)$ <u>over</u> F <u>corresponding to the first</u> ℓ-<u>adic cohomology groups</u> $H^1(A_f^i \times \overline{F}, \mathbb{Q}_\ell)$, <u>is given by the product of</u> L-<u>functions</u>

$$\prod_{\sigma: K_f \hookrightarrow \mathbb{C}} L^{(1)}(s, f^\sigma/F).$$

<u>Remark</u> 3. This conjecture is also true for selfconjugate forms f, because $L^{(1)}(s,f/F)=L(s,h)L(s,h^\rho)$, if $f=DN_Q(h)$ (cf. §12.4), and because the product

$$\prod_{\sigma: K_h \hookrightarrow \mathbb{C}} L(s,h^\sigma)$$

gives L-function of the $Gal(\overline{F}/F)$-module $H^1_{ét}(B_h \times \overline{F}, \mathbb{Q}_\ell)$ (cf. Shimura [43] Chap.7).

There is an access to these two conjectures for primitve forms of higher level, even if they are not selfconjugate. I shall discuss this in subsequent papers.

20.3. Let us consider the $Gal(\overline{F}/F)$-module $H^2(M_f, \mathbb{Q}_\ell)$ defined in the previous section for any primitive form f of $S_2(SL_2(O_F))$.

By Asai [2], we know the relation between the L-function $L^{(2)}(s,f/\mathbb{Q})$ and the L-function $L^{(1)}(s,f/F)$, and especially for any selfconjugate form $f=DN_Q(h)$, the relation between $L^{(2)}(s,f/\mathbb{Q})$ and the convolution $L(s, h, h^\rho)$ (cf. Ogg [34]). On the other hand, we have an isomorphism of K_f-Hodge structures (Main Theorem A of §7)

$$H^2(M_f, \mathbb{Q}) \cong H^1(A_f^1, \mathbb{Q}) \otimes_{K_f} H^1(A_f^2, \mathbb{Q}),$$

and especially for selfconjugate form $f=DN_Q(h)$ with a primitive form h of $S_2(\Gamma_0(D), \varepsilon_D)$, an isomorphism of $K_f=k_h$-Hodge structures (Main Theorem B of §17)

$$H^2(M_f, \mathbb{Q}) \cong H^1(B_h, \mathbb{Q}) \otimes_{k_h} H^1(B_h, \mathbb{Q}).$$

In view of the above relation of L-functions and the above isomorphisms of K_f-Hodge structures, we have the following conjecture on ℓ-adic cohomology, by the "alchemy of motives" in Deligne [12].

<u>Conjecture</u> 3. (ℓ-<u>adic realization</u>). <u>For any primitive form</u> f <u>of weight</u> 2, <u>there is an isomorphism of</u> $Gal(\overline{F}/F)$-$(K_f \times \mathbb{Q}_\ell)$-<u>bimodules</u>

$$H^2(M_f, \mathbb{Q}_\ell) \cong H^1(A_f^1 \times \overline{F}, \mathbb{Q}_\ell) \otimes_{K_f} H^1(A_f^2 \times \overline{F}, \mathbb{Q}_\ell).$$

<u>Especially if</u> $f=DN_Q(h)$ <u>for some primitive form</u> h <u>of</u> $S_2(\Gamma_0(D), \varepsilon_D)$, <u>there is an isomorphism of</u> $Gal(\overline{F}/F)$-$(K_f \times \mathbb{Q}_\ell)$-<u>bimodules</u>

$$H^2(M_f, \mathbb{Q}_\ell) \cong H^1(B_h \times \overline{F}, \mathbb{Q}_\ell) \otimes_{k_h} H^1(B_h \times \overline{F}, \mathbb{Q}_\ell),$$

<u>Note here that</u> $k_h=K_f$.

Remark 4. The example of §19.3 is an evidence for Conjecture 3. This conjecture seems to be difficult to prove. However, the latter part of Conjeture 3 follows by Čebotarev's density theorem, if we know the L-function of $H^2(M_f,\mathbb{Q}_\ell)$ as claimed in [8], and if we know the semi-simplicity of the actions of the Frobenius elements in $Gal(\overline{F}/F)$ on the module $H^2(M_f,\mathbb{Q}_\ell)$. Because we can check the coincidence of the characteristic polynomials of the Frobenius mappings of the both hand sides of Conjecture 3.

Acknowledgement. I would like to thank Momose, K. Murthy and Ribet for discussing Conjecture 3 with me.

Bibliography.

[1] Artin,M.:Comparison avec la cohomologie classique (exposé XI), Théorie des topos et cohomologie étale des schémas (SGA 4), t.3, pp.64-78. Lect. Notes in Math., No.305. Springer 1973.

[2] Asai,T.:On certain Dirichlet series associated with Hilbert modular forms and Rankin's method. Math. Ann. 226, 81-94 (1977).

[3] Asai,T.:On the Doi-Naganuma lifting associated with imaginary quadratic fields. Nagoya Math. J. 71, 149-167 (1978).

[4] Baily,W.L.& Borel,A.:Compactification of arithmetic quotients of bounded symmetric domains. Annals of Math. 84, 442-528 (1966).

[5] Birch,B.J.& Swinnerton-Dyer,H.P.F.:Notes on elliptic curves I; II. Journ. f. reine u. angew. Math. 212, 7-25 (1963);218, 79-108 (1965).

[6] Birch,B.J.:Elliptic curves over $\mathbb{Q}$; a progress report. Number Theory Institute 1969. Proc. Symp. Pure Math. 20. Amer. Math. Soc. 369-400 (1971).

[7] Bourbaki,N.:Algèbre. Hermann;Paris (Act. Sci. et Ind. 1272) 1959.

[8] Casselmann,W.:The Hasse-Weil ζ-function of some moduli varieties of dimension greater than one. Automorphic forms, representations, and L-functions. A.M.S. Proc. of symp. pure math. 33. Part 2, 141-163 (1979).

[9] Deligne,P.:Formes modulaires et representations ℓ-adiques. Seminaire Bourbaki, 1968/1969, expose n^0 355. 139-172. Lecture Notes in Math., 179. Springer. 1971.

[10] Deligne,P.:La conjecture de Weil pour les surfaces K3. Invent. math. 15, 206-226 (1972).

[11] Deligne,P.:Theorie de Hodge II ;III. Pub. Math. Inst. Hautes Etudes Sci., 40, 5-58 (1971);44, 5-77 (1974).

[12] Deligne,P.:Valuers de fonctions L et periodes d'integrales. Automorphic forms, representations, and L-functions. A.M.S. Proc. of symp. pure math. 33. Part 2, 313-346 (1979).

[13] Doi,K.& Naganuma,H.:On the functional equation of certain Dirichlet series. Invent. math. 9, 1-14 (1969).

[14] Harder,G.:On the cohomology of SL(2,$\mathcal{O}$). Lie groups and their representations, 139-150. (Proc. of the summer school on group representations. Budapest. 1971) Adam Hilger; London. 1975.

[15] Hida,H.:On the abelian varieties with complex multiplication as factors of the abelian variety attached to Hilbert modular forms. Japan j. Math. 5, 157-208 (1979).

[16] Hirzebruch,F.:Hilbert modular surfaces. L'Ens. Math. 19, 183-281 (1973).

[17] Hirzebruch,F.:Kurven auf den Hilbertschen Modulflächen und Klassenzahlrelationen. Classification of algebraic varieties and compact manifolds. Lecture Notes in Math. 412, pp.75-93. Springer; Berlin-Heidelberg-New York. 1974.

[18] Hirzebruch,F.& Van de Ven.:Hilbert modular surfaces and the classification of algebraic surfaces. Invent. math. 23, 1-29 (1974).

[19] Hirzebruch,F.& Zagier,D.:Intersection numbers of curves on Hilbert modular surfaces and modular forms of Nebentypus. Invent. math. 36, 57-113 (1976).

[20] Kuga,M.& Satake,I.:Abelian varieties attached to polarized K3 surfaces. Math. Annalen 169, 239-242 (1967); Corrections. ibid. 173, 322 (1967).

[21] Langlands,R.P.:On the zeta-functions of some simple Shimura varieties. Cand. J. of Math. 31. No.6, 1121-1216 (1979).

[22] Manin,Y.I.:Parabolic points and zeta functions of modular curves. Izv. Akad. Nauk SSSR. Ser. Mat. 36, 19-66 (1972).

[23] Mazur,B.& Swinnerton-Dyer,P.:Arithmetic of Weil curves. Invent. math. 25, 1-61 (1974).

[24] Miyake,T.:On automorphic forms on GL_2 and Hecke operators. Ann. of Math. 94, 174-189 (1971).

[25] Momose,F.:On the ℓ-adic representations attached to modular forms. J. of Fac. Sci. Univ. of Tokyo. (Sec. IA) 28, 89-109 (1981).

[26] Mountjoy,R.H.:Abelian varieties attached to representations of discontinuous groups. Amer. J. of Math. 89, 149-224 (1967).

[27] Mumford,D.:Abelian varieties. Tata Inst. of Fund. Research Studies in Math. Oxford Univ. Press. 1970.

[28] Mumford,D.:Hirzebruch's proportionality theorem in the non-compact case. Invent. math. 42, 239-272 (1977).

[29] Naganuma,H.:On the coincidence of two Dirichlet series associated with cusp forms of Hecke's "Neben"-type and Hilbert modular forms over real quadratic field. J. Math. Soc. Japan 25, 547-555 (1973).

[30] Niwa,S.:Modular forms of half integral weight and the integral of certain theta-functions. Nagoya Math. J. 56, 147-161 (1974).

[31] Oda,T.:On modular forms associated with indefinite quadratic forms of signature (2,n-2). Math. Ann. 231, 97-144 (1977).

[32] Oda,T.:A note on the Tate conjecture for K3 surfaces. Proc. of Japan Acad. 56. Ser.A, No.6, 296-300 (1980).

[33] Oda,T.:On the poles of Andrianov L-functions. Math. Annalen 256, 323-340 (1981).

[34] Ogg,A.P.:On a convolution of L-series. Inventiones math.7, 297-312 (1969).

[35] Rallis,S.& Schiffmann,G.:On a relation between $\widetilde{SL}_2$ cusp forms and cusp forms on tube domains associated to orthogonal groups. Trans. Amer. Math. Soc. 263, 1-58 (1981).

[36] Rapoport,M.:Compactification de l'espace de modules de Hilbert-Blumenthal. Compositio math. 36, 255-335 (1978).

[37] Ribet,K.A.:Twists of modular forms and endomorphisms of abelian varieties. Math. Annalen 253, 43-62 (1980).

[38] Saito,H.:Automorphic forms and algebraic extensions of number fields. Lectures in mathematics. Kinokuniya-Bookstore;Tokyo. 1975.

[39] Saito,M.:Representations unitaires des groupes symplectiques. J. Math. Soc. of Japan 24, 232-251 (1972).

[40] Satake,I.:Clifford algebras and families of abelian varieties. Nagoya Math. J. 27-2, 435-446 (1966). Corrections. ibid. 31, 295-296 (1968).

[41] Shimura,G.:Sur les integrales attachees aux formes automorphes. J. Math. Soc. of Japan 11, 291-311 (1959).

[42] Shimura,G.:On analytic families of polarized abelian varieties and automorphic functions. Ann. of Math. 78, 149-192 (1963).

[43] Shimura,G.:Introduction to the arithmetic theory of automorphic functions. Iwanami Shoten (Tokyo) and Princeton Univ. Press. 1971.

[44] Shimura,G.:On elliptic curves with complex multiplication as factors of the jacobian of modular function fields. Nagoya Math. J. 43, 199-208 (1971).

[45] Shimura,G.:Class fields over real quadratic fields and Hecke operators. Ann. of Math. 95, 130-190 (1972).

[46] Shimura,G.:On canonical models of arithmetic quotients of bounded symmetric domains I; II. Ann. of Math. 91, 144-222 (1970);92, 528-549 (1970).

[47] Shimura,G.:The special values of zeta functions associated with cusp forms. Communications on pure and applied math. 29, 783-804 (1976).

[48] Shimura,G.:On the periods of modular forms. Math. Annalen 229, 211-221 (1977).

[49] Shintani,T.:On construction of holomorphic cusp forms of half integral weight. Nagoya Math. J. 58, 83-126 (1975).

[50] Siegel,C.L.:Lectures on Advanced Analytic Number Theory. Tata Institute of Fundamental Research;Bombay 1961 (revised, 1965).

[51] Siegel,C.L.:Indefinite quadratische Formen und Funktionentheorie I. Math. Annalen 124, 17-54 (1951)=Gesam. Abh. Bd III.58 (Springer Verlag), 105-142, 1966.

[52] van der Geer,G.& Ueno,K.:Families of abelian surfaces with real multiplication over Hilbert modular surfaces. To appear in Nagoya Math. J. (1982)

[53] Weil,A.:Sur certains groupes d'operateurs unitaires. Acta math. 111, 143-211 (1964).

[54] Weil,A.:Über die Bestimmung Dirichletscher Reihen durch Funktionalgleichungen. Math. Annalen 168, 149-156 (1967).

[55] Zagier,D.:Modular forms associated to real quadratic fields. Inventiones math. 30, 1-46 (1975).

[56] Zagier,D.:Modular forms whose Fourier coefficients involve zeta-functions of quadratic fields. Modular functions of one variable VI, 105-169. Lecture Notes in Math. no.627. Springer;Berlin-New York-Heidelberg. 1977.

[57] Jacquet,H.& Shalika,J.A.:A non-vanishing theorem for zeta function of GL_n. Inventones math. 38, 1-16 (1976).

[58] Rapoport,M.:Complément à l'article de P. Deligne «La conjecture de Weil pour les surfaces K3». Inventiones math.15, 227-236 (1972)

[59] Oda,T.:Periods of Hilbert modular surfaces. Proc. of Japan Acad. 57, Ser A (8), 415-419 (1981).

[60] Shioda,T.:On elliptic modular surfaces. J. Math. Soc. Japan 24, 20-59 (1972)

[61] Tate,J.:Algebraic cycles and poles of zeta functions. Arithmetic algebraic geometry. New York;Harper & Row. 1966.

[62] Hirzebruch, F. Modulflähen und Modulkurven zur symmetrichen Hilbertschen Modelgruppe. Ann. Scient. Ec. Norm. Sup. 11, 101-166 (1978).

Progress in Mathematics
Edited by J. Coates and S. Helgason

Progress in Physics
Edited by A. Jaffe and D. Ruelle

- A collection of research-oriented monographs, reports, notes arising from lectures or seminars
- Quickly published concurrent with research
- Easily accessible through international distribution facilities
- Reasonably priced
- Reporting research developments combining original results with an expository treatment of the particular subject area
- A contribution to the international scientific community: for colleagues and for graduate students who are seeking current information and directions in their graduate and post-graduate work.

Manuscripts

Manuscripts should be no less than 100 and preferably no more than 500 pages in length.

They are reproduced by a photographic process and therefore must be typed with extreme care. Symbols not on the typewriter should be inserted by hand in indelible black ink. Corrections to the typescript should be made by pasting in the new text or painting out errors with white correction fluid.

The typescript is reduced slightly (75%) in size during reproduction; best results will not be obtained unless the text on any one page is kept within the overall limit of 6 × 9½ in (16 × 24 cm). On request, the publisher will supply special paper with the typing area outlined.

Manuscripts should be sent to the editors or directly to:
Birkhäuser Boston, Inc., P.O. Box 2007, Cambridge, Massachusetts 02139

PROGRESS IN MATHEMATICS
Already published

PM 1 Quadratic Forms in Infinite-Dimensional Vector Spaces
Herbert Gross
ISBN 3-7643-1111-8, 431 pages, paperback

PM 2 Singularités des systèmes différentiels de Gauss-Manin
Frédéric Pham
ISBN 3-7643-3002-3, 339 pages, paperback

PM 3 Vector Bundles on Complex Projective Spaces
C. Okonek, M. Schneider, H. Spindler
ISBN 3-7643-3000-7, 389 pages, paperback

PM 4 Complex Approximation, Proceedings, Quebec, Canada, July 3-8, 1978
Edited by Bernard Aupetit
ISBN 3-7643-3004-X, 128 pages, paperback

PM 5 The Radon Transform
Sigurdur Helgason
ISBN 3-7643-3006-6, 202 pages, paperback

PM 6 The Weil Representation, Maslov Index and Theta Series
Gérard Lion, Michèle Vergne
ISBN 3-7643-3007-4, 345 pages, paperback

PM 7 Vector Bundles and Differential Equations
Proceedings, Nice, France, June 12-17, 1979
Edited by André Hirschowitz
ISBN 3-7643-3022-8, 255 pages, paperback

PM 8 Dynamical Systems, C.I.M.E. Lectures, Bressanone, Italy, June 1978
John Guckenheimer, Jürgen Moser, Sheldon E. Newhouse
ISBN 3-7643-3024-4, 300 pages, paperback

PM 9 Linear Algebraic Groups
T. A. Springer
ISBN 3-7643-3029-5, 304 pages, hardcover

PM10 Ergodic Theory and Dynamical Systems I
A. Katok
ISBN 3-7643-3036-8, 352 pages, hardcover

PM11 18th Scandinavian Congress of Mathematicians, Aarhus, Denmark, 1980
Edited by Erik Balslev
ISBN 3-7643-3034-6, 528 pages, hardcover

PM12 Séminaire de Théorie des Nombres, Paris 1979–80
Edited by Marie-José Bertin
ISBN 3-7643-3035-X, 408 pages, hardcover

PM13 Topics in Harmonic Analysis on Homogeneous Spaces
Sigurdur Helgason
ISBN 3-7643-3051-1, 142 pages, hardcover

PM14 Manifolds and Lie Groups, Papers in Honor of Yozô Matsushima
Edited by J. Hano, A. Marimoto, S. Murakami, K. Okamoto, and H. Ozeki
ISBN 3-7643-3053-8, 480 pages, hardcover

PM15 Representations of Real Reductive Lie Groups
David A. Vogan, Jr.
ISBN 3-7643-3037-6, 771 pages, hardcover

PM16 Rational Homotopy Theory and Differential Forms
Phillip A. Griffiths, John W. Morgan
ISBN 3-7643-3041-4, 264 pages, hardcover

PM17 Triangular Products of Group Representations and their Applications
S. M. Vovsi
ISBN 3-7643-3062-7, 150 pages, hardcover

PM18 Géométrie Analytique Rigide et Applications
Jean Fresnel, Marius van der Put
ISBN 3-7643-3069-4, 215 pages, hardcover

PROGRESS IN PHYSICS
Already published

PPh1 Iterated Maps on the Interval as Dynamical Systems
Pierre Collet and Jean-Pierre Eckmann
ISBN 3-7643-3026-0, 256 pages, hardcover

PPh2 Vortices and Monopoles, Structure of Static Gauge Theories
Arthur Jaffe and Clifford Taubes
ISBN 3-7643-3025-2, 275 pages, hardcover

PPh3 Mathematics and Physics
Yu. I. Manin
ISBN 3-7643-3027-9, 111 pages, hardcover